# *Laboratory Manual*

*to accompany*

Schauf / Moffett / Moffett

# *Human Physiology*
## *Foundations & Frontiers*

B. J. WEDDELL                    DAVID F. MOFFETT

**TIMES MIRROR/MOSBY COLLEGE PUBLISHING**

*St. Louis · Boston · Toronto*

*Publisher:* Edward F. Murphy
*Editor:* Deborah Allen
*Developmental Editor:* Jean Babrick
*Book Designer:* Sue Pepe
*Original Illustrations:* Eileen Draper
*Camera Ready Production:* Pepe Productions

*Illustration credits:*

2-1, 2-2, Ronald J. Ervin
3-1A, William Ober
3-1B, Brenda R. Eisenberg, Ph.D., University of Illinois at Chicago
4-4, Ronald J. Ervin
5-1, William Ober
5-3, Barbara Stackhouse
8-4, Joan M. Beck/Andrew Grivas
8-5, Joan M. Beck
10-2, 10-4, Barbara Stackhouse
10-3, Brenda R. Eisenberg, Ph.D., University of Illinois at Chicago
10-5A, G. David Brown
11-1, William Ober
11-2A, Marsha J. Dohrmann
11-2B, E.R. Lewis/F.S. Werblin/Y.Y. Feevi, University of California, Berkeley
11-4, 11-5, G. David Brown
12-1, 12-3A, Marsha J. Dohrmann
12-2, Barbara Stackhouse
12-3B, C, Kathy Mitchell Grey
12-4, G. David Brown
12-6, Terry Cockerham, Synapse Media Production
14-1, Dr. Andrew P. Evan, Indiana University
14-2, Brenda R. Eisenberg, Ph.D., University of Illinois at Chicago
14-3, Michael P. Schenk
14-4, 14-5, Barbara Stackhouse
14-6, Dr. Andrew P. Evan, Indiana University
15-3, 15-4, Barbara Stackhouse
16-2, Rusty Jones
16-3, Barbara Stackhouse
17-1, 17-4, 17-5, Barbara Stackhouse
18-1, Joan M. Beck/Donna Odle
18-2, 18-3, 18-4, Barbara Stackhouse
19-1A, 19-2A, 19-3, 19-6, Barbara Stackhouse
19-1B, Patricia Kane, Indiana University Medical Center, Radiology Department
20-1B, Brenda R. Eisenberg, Ph.D., University of Illinois at Chicago

*T*his laboratory manual is based on experience gained in teaching an introductory course in Human Physiology given by faculty of the Program in Zoophysiology at Washington State University. In addition to standard laboratory exercises, the manual contains special computer exercises designed to accompany commercially available software. Selected exercises are also available to qualified adopters.

Each exercise has been refined over numerous semesters. The necessary equipment is generally available in teaching laboratories. The accompanying Preparator's Manual includes answers to questions in the laboratory manual, recipes for solutions, information about setting up equipment, suggestions for directing student laboratory reports, a list of available software for the computer exercises, addresses of equipment suppliers, and guidelines and tips based on our experience with these exercises.

**This manual is fully integrated with the new Schauf, Moffett, and Moffett Human Physiology: Foundations and Frontiers.** Each exercise lists relevant pages in the text for preparation before the lab and as a study aid afterwards. The sequence of exercises parallels the order of topics in Schauf, Moffett, and Moffett and suits the syllabus most instructors use.

We had the following objectives:

**To examine in depth the concepts illustrated in the exercises**
Some laboratory manuals are simply lists of steps to be carried out. This "cookbook" approach makes it easy for students to stop thinking and lapse into following directions. This book in contrast, provides detailed background information including relevant illustrations. Important facts, concepts, and terms are reinforced throughout the exercises.

**To develop an understanding of the scientific method**
Many exercises begin with stated hypotheses that are to be tested or with hypotheses that students complete. In addition, students may be asked to predict results for the experiment they are about to perform. This approach helps the student to formulate expectations before doing the exercises, so that the results can be evaluated and understood in terms of these expectations.

**To stimulate thinking in the laboratory setting**
While the laboratory experience is still fresh, students enter their data and answer questions that tie the laboratory experience to what they have learned from lectures and reading assignments.

## To provide students with opportunities to observe their own physiology

Most students find that it is easy to be interested in the subject when the subject is their own body. In many of these exercises, students themselves are the subjects. For example, students make and interpret their own electrocardiograms, measure their cardiovascular and respiratory responses to exercise, measure blood pressure and vital capacity, observe their reflex responsiveness, and determine their blood types and sensory thresholds.

## To familiarize students with the methods of physiological and medical research

As well as experimenting on themselves, students carry out exercises involving experimental animals and biochemical reactions. In exercises that require nonhuman subjects, such as the skeletal muscle, intestinal smooth muscle, heart, and endocrine experiments, the relevance to human physiology is made clear. In Exercise 21 students gain experience in experimenting on living animals over several weeks and performing clinical procedures, such as giving injections.

## To develop the students' skills

Instructors know that "students learn by doing," and nowhere is this more true than in the laboratory setting. We have designed this manual to develop the students' skills by giving them practice in the following areas:

### Compiling data

In each exercise, data are entered in tables using the format of a scientific journal.

### Handling quantitative data

Students work with quantitative data in most exercises in this manual. In some, they perform calculations using their data; in others, they plot data. Explicit instructions are given to guide students who are inexperienced or uncomfortable handling quantitative data.

### Writing laboratory reports

In addition to the laboratory report pages at the end of each exercise, we recommend that students submit written reports on certain exercises. This gives them an opportunity to become familiar with the proper format for presenting and discussing scientific data. It also gives them a chance to practice written communication of what they have learned. Appendix 3 provides the student with guidelines for preparing written reports.

## Using physiological instruments

Students use a variety of instruments in these exercises. The use of each instrument is explained and illustrated the first time it is introduced.

## Using computers

Three exercises allow students to simulate experiments with a nerve axon, a skeletal muscle, and the cardiovascular system. The programs could be used to replace the corresponding hands-on exercises but are more effective when used in conjunction with them. They are affordable, readily available, and run on standard IBM personal computers or compatibles.

## ACKNOWLEDGEMENTS

We would like to acknowledge the contributions of our colleagues, Drs. Leonard Kirschner, Alan Koch, Stacia Moffett, and Hans Went to the development of these exercises. We also thank Jean Babrick for editorial assistance, Dennis Cartwright for useful advice regarding the exercises, Eileen Draper for preparing new artwork, Joan Folwell for technical assistance, Sue Pepe for assistance with design, and all of the above for their patience as we revised this project. Helpful reviews of the manuscript were provided by Dr. Louis Burnett, University of San Diego; Dr. Keith Dupre, University of Nevada at Las Vegas; Dr. John Scheide, Central Michigan University; and Dr. Gregory Stephens, University of Delaware. Finally, we are indebted to the hundreds of Zoology 251 students who provided their feedback and helped us to refine these exercises.

B.J. WEDDELL
DAVID F. MOFFETT

## *Physiology: An Experimental Science*

There are many sources of information about how the human body works. These include observations on human patients with diseases and injuries as well as experiments on healthy human subjects, animals, and cultured cells and tissues. Experiments in physiology typically involve making comparisons between a set of **control** data and one or more **experimental** data sets. Ideally, the experimental data sets reflect the effects of a single variable that is under the control of the experimenters.

One might compare a group of patients receiving one treatment with a control group receiving no treatment. In such an experiment, every effort is made to ensure that the patients in the two groups are similar in every important respect except the treatment, so that any differences between the two groups can be attributed to the treatment. For example, each patient in the treatment group might be matched by age, sex, and other variables to a patient in the control group. This is the approach you will use in Exercise 21 on the effects of testosterone on growth and development.

In some experiments, each subject serves as a control and receives one or more experimental treatments as well. For example, in Exercise 15 you will determine the effects of various drugs and hormones on the rhythm of the heart. Before and after each application of a test substance you will wash the heart and let it beat for a time to observe its rate in the absence of outside influences. These control intervals serve as the basis for detecting and measuring the effects of the test substances.

A **hypothesis** is a statement that can be tested by experiment. In many cases hypotheses are statements about cause-and-effect relationships. The goal of experiments is to test hypotheses.

For example, the Canadian scientists Bayliss and Starling formulated the following hypotheses: 1) the pancreas releases a hormone that allows cells to take up glucose from the blood; 2) a lack of this hormone causes the disease diabetes mellitus, characterized by very high blood glucose levels. To test their hypotheses, they removed the pancreases of dogs. The dogs then showed the symptoms of diabetes mellitus, supporting the hypotheses. In contemporary terms we would call this surgery-induced diabetes an **animal model** of the human disease. Additional support for the hypothesis was provided when Bayliss and Starling found that injection of crude extracts of pancreatic tissue relieved the diabetes of their subjects. After showing that they could reverse the diabetes caused by surgical removal of the pancreas in dogs, the next step was to show that the same extracts could be effective against the human disease. Exercise 21 on the effects of testosterone on growth and sexual differentiation in chicks is conceptually similar to the experiments of Bayliss and Starling.

If the same experiment is performed several times, the results may differ substantially. This variation is called **experimental error**. The term is confusing, because it suggests that the variability of the outcome is the result of experimenters' mistakes in carrying out the experiment. Actually experimental error does not include the effects of experimenters' mistakes. Rather, it is a result of the operation of chance on the measuring instruments and the animals or tissues that are the subjects of the experiment.

Variation in experimental results occurs for several reasons:

**First, the act of measuring introduces error.** For example, if the same sample is weighed five times on the same sensitive balance, a different weight will be obtained each time. The smaller the range of different weights obtained, the higher the **precision** of the balance. Obviously, if the same sample is weighed a large number of times and the weights are averaged, errors of imprecision will tend to cancel out. If few weighings, or just one, are used, chance will make the measurement much less certain. In contrast to precision, the **accuracy** of the balance is the degree to which its weights would match a recognized standard. If the balance were not well calibrated, its results would be wrong, even if it is a precision instrument. Such an "error" of inaccuracy is not due to experimental error.

**Biological systems themselves are innately variable.** If we wanted to know if diet had a significant effect on height, we might compare the heights of two different ethnic groups with different diets. In this example, dietary differences are the "treatment." However, when the variability among subjects is large, the average values for a measured variable may differ between control and experimental groups as a result of chance alone, even if the treatment really has no effect. Thus, even if we found differences in height between the two different ethnic groups, differences in diet might not have caused the differences in height.

The likelihood that the differences in height were due to the treatment can be assessed with statistical methods. Scientists use statistical techniques to calculate the probability that differences between experimental and control data sets are due to chance alone. If the probability that the differences are due to chance alone is low, then we can conclude that it is likely that the differences are caused by the experimental treatment. Although such techniques are beyond the scope of this course, you should recognize that, even in well-controlled experiments, the effects of chance and innate biological variation make it necessary to perform a number of replications before it is possible to draw valid conclusions about the outcome of the experiment. This is one reason why physiological research projects sometimes involve many animals.

One source of variability between individual animals or people is genetic. In some cases it is possible to overcome this variability by studying identical twins or highly inbred animals. Another type of variability is state-dependent. The body alternates between several recognizable physiological states, such as sleeping and waking, or the absorptive state that immediately follows a meal and the postabsorptive state that begins several hours after one's last meal. Variability due to these sources can be reduced by controlling the subjects' environment over time. For example, in the experiment on body fluid homeostasis (Exercise 19) it would be desirable to have all subjects fast for a time to eliminate the effects of eating and drinking in the hours preceding the experiment.

"My experiment didn't work!" is a lament heard innumerable times in student laboratories. This means that the outcome didn't match what students were led to expect by their textbook or instructor. There are several possible causes of these "failures," and they are all potentially instructive. For instance, the experimenter may have made an error or the equipment may have failed. Tracing out such a failure, perhaps with the aid of your instructor, will help you to learn lab habits and techniques that guard against mistakes and to gain a better understanding of how your equipment works. Sometimes "failure" reflects the biology of the animal. For example, the responsiveness of tissues of "cold-blooded" animals such as the turtle and frog is affected by the season of the year and the temperature at which the animals are held before use.

In addition, you should be alert to "failures" that may reveal something new. For example, a classical student experiment in immunology shows that animals injected with foreign substances begin to produce molecules called antibodies that bind specifically to the injected substances. An important insight about antibody production came from researchers studying the embryological effects of removing a structure called the bursa of Fabricius from chick embryos. When the experiments were completed, the mature chickens were used in a student laboratory on antibody production. However, the students found they could not get their birds to make antibodies; they thought at first that their experiment had "failed." But, follow-up experiments showed that in birds cells that will later make antibodies must spend some time in the bursa of Fabricius during early development. The experimental chickens had failed to make antibodies because their bursae had been removed. Today we call these antibody-producing cells B cells; the "B" stands for bursa. (We still do not know exactly which areas of the mammalian body correspond functionally to the bursa.) In this example, the "failed" student experiments led to an important discovery.

There are many other examples of breakthroughs that resulted from experimental results that "failed" to conform to expectations. The woman who first correctly determined the number of chromosomes in human cells was discouraged because her results did not agree with those of her colleagues. It turned out, however, that she was right and they had counted the chromosomes incorrectly. These examples underscore the importance of integrity. No matter what results you obtain in your laboratory exercises, it is essential that you report them honestly.

## The Use of Computers in Physiology

A **mathematical model** is an attempt to duplicate the behavior of a natural system using mathematical relationships between the important variables of the system. The availability of computers has made such models useful, because computers can carry out the calculations much more rapidly than they could be done by hand. For example, in the model of skeletal muscle included with this manual, the force developed by a stimulated muscle is a mathematical function of the length of the muscle, the stimulus frequency, the stimulus intensity, and the duration over which the muscle is stimulated. Although it would be extremely tedious to keep track of all these variables without computers, you can use this model without doing a single calculation.

When you use the computer models included in this manual, remember that, compared to the real systems, these models are greatly simplified. A more complete model of skeletal muscle would include additional variables, such as temperature, the composition of the extracellular fluid, or the effects of hormonal inputs.

Computer-based models are increasingly important in research, because if a model's behavior resembles that of the real system in all points, the model probably incorporates correct statements about the relationships between the variables of the system. Since the computer model does not incorporate any characteristics except those it is programmed for, computer models are not sources of "new" information about physiology, but they can provide useful insights into how physiological systems behave.

The validity of computer models must be tested continually by comparing their behavior to that of the system being modeled. Thus, computer models cannot be thought of as substitutes for animal experiments in research. However, they can be valuable complements to the animal experiments you will carry out.

## The Use of Animals in Physiology

Physiology is a way of knowing about how the body works. In physiology the basis of knowledge is experimentation. For some experiments in this manual, you will utilize your own body and the bodies of your classmates. For others, you will use the bodies of animals or fresh, living tissue. This is appropriate, because most of what we know about how the human body works is derived from experiments on animals. To the student of physiology, animal experimentation is as important as carrying out chemical reactions is to the chemistry student. However, these experiments sometimes require the death of the animal involved.

There are three important things to keep in mind in such experiments:

**First, suffering will be minimized.** In these experiments, your instructor will rapidly destroy the higher parts of the central nervous system of the animal. This will be done mechanically or with an overdose of anesthesia. Many processes in the body can continue for a time in the absence of a central nervous system. For instance, reflexes that are mediated by the spinal cord continue even though the animal no longer feels any sensation. This makes it possible for you to observe physiological processes operating in living tissues, while the experimental animals will feel no pain.

**Secondly, you should show respect for the animal that has been sacrificed by learning as much as you can from each animal experiment.** This calls for care and organization in carrying out the experiments and in analyzing the data.

**Finally, animal experiments allow students to experience the beauty and fascination of working organs in a living animal.** There is no substitute for experiencing this esthetic appreciation directly. A beating heart that speeds up or slows down as you give it neural and hormonal stimulation, or a tiny muscle that valiantly lifts its load when you stimulate it, are no less remarkable now than they were when these experiments were first carried out. It is our hope that you will gain an appreciation of these processes as you do the experiments in this manual and that these exercises will stimulate your sense of wonder at the workings of living creatures.

If you approach these exercises with the proper attitude, we believe that the sacrifice of animal life involved will be justified by the gains in knowledge, respect, and appreciation that result.

# Table of Contents

# Effect of Temperature on Chemical Reactions in Cells

*Reading assignment: text 28, 34-35*

## Objectives

### Experimental

1. To determine the effect of ambient temperature on the ventilation rate of a frog.

2. To determine the effect of ambient temperature on the heart rate of a) a tadpole and b) a frog.

### Conceptual

After completing this exercise and the reading assignment, you should be able to:

1. Describe the effects of a) increasing and b) decreasing ambient temperature on the ventilation rate of a frog and explain why temperature has these effects.

2. Describe the effects of a) increasing and b) decreasing ambient temperature on the heart rate of tadpoles and frogs and explain why temperature has these effects.

3. Define the terms **kinetic energy, activation energy, ambient temperature, homeotherm, poikilotherm, myogenic,** and **denaturation.**

4. Explain how changes in temperature affect the rates of chemical reactions.

5. Give one reason why there is an upper limit to the temperatures that living organisms can tolerate.

6. Explain how a temperature quotient is calculated.

## Background

Temperature is one of the most important factors in determining the rate of chemical reactions, including reactions taking place in living systems. The more a substance is heated, the more its molecules vibrate. The energy of motion in the vibrating molecules is **kinetic energy**. Because **kinetic energy** increases as temperature increases, the rate at which molecules of reactants collide with each other is also a function of temperature.

For example, hydrogen and oxygen can combine to produce water, according to the following equation:

$$2H_2 + O_2 \rightarrow 2H_2O.$$

At room temperature, this reaction proceeds very slowly. However, if a mixture of hydrogen and oxygen is heated, the rate of water formation increases markedly.

In any chemical reaction, the molecules of the reactants must be supplied with a quantity of energy that is sufficient to make or break chemical bonds. This quantity of energy is the **activation energy** for the reaction. When reactants are heated, molecules are more likely to collide with enough kinetic energy to react.

Like other chemical reactions, the reactions going on in the human body are affected by temperature. Consequently many physiological processes depend on temperature. Within certain limits (see below), as temperature increases, the rate of cellular metabolism increases. As a consequence of this accelerated metabolic rate, oxygen demand also rises.

To meet the increased demand for oxygen, the rates of physiological processes involved in obtaining oxygen and delivering it to the tissues also increase. Thus, under certain circumstances, heart rate and breathing rate would be expected to rise as temperature increases. If we could run the reactions of cellular metabolism in a test tube at different temperatures, we would be able to observe this temperature dependence directly.

Another approach to demonstrating this phenomenon would be to take an organism, vary its internal temperature, and then measure the rates of cellular reactions at different temperatures. However, this experimental design will only work if we have a way of controlling the internal temperature of our experimental animal.

Birds and mammals, including human beings, are **homeotherms**, that is, they physiologically regulate their internal temperature within a narrow range. They do this, within limits, regardless of the temperature of their surroundings (**ambient temperature**). Obviously, then, our proposed experimental design will not work with humans or other homeotherms.

Fortunately, there are other animals that fulfill the requirements of our experiment. Animals whose body temperature more or less parallels ambient temperature are called **poikilotherms** and are commonly, but erroneously, referred to as "cold-blooded." Invertebrates, fishes, amphibians, and reptiles are poikilothermic.

In this experiment, you will determine the rate of two physiological processes in a poikilothermic animal. You will measure the rate at which a frog lowers the floor of its mouth to take in air (**ventilation rate**). You will also observe the beating hearts of tadpoles and frogs at a variety of ambient temperatures. Because the body of young tad-

poles of the genus *Xenopus* is translucent, the beating heart can be readily observed with a dissecting microscope.

To observe the response of heart rate to changing temperature in an adult frog, you will take advantage of the fact that a frog's heart continues to beat for some time after it is removed from the body. This occurs because the vertebrate heart is **myogenic**, that is, contraction originates spontaneously in the heart muscle. The organs of most vertebrates will survive and function for many hours, even days, if they are immersed in a Ringer's solution which is buffered at pH 7 and aerated. (Ringer's is a solution with about the same salt composition as body fluids.) This is true for human tissue, although human tissue is less hardy in this respect.

Before beginning your experiment, it is useful to have a mental model of how the experimental system behaves. It is helpful to summarize such a model in the form of one or more hypotheses (singular: hypothesis). Your experiments should be designed to test your hypotheses.

> **Hypothesis 1:** Within certain limits, the rates of physiological processes involved in obtaining oxygen and delivering it to the tissues increase as the temperature of a system increases.

To be sure you understand the implications of this hypothesis, complete the predictions below.

> **Prediction 1:** Ventilation rate and heart rate in a poikilothermic animal will
>
> _____
>
> when ambient temperature increases.

## Prediction 2: Ventilation rate and heart rate in a poikilothermic animal will

_____

when ambient temperature decreases.

As a general rule, metabolic rates in poikilotherms are about two to three times higher for every $10°C$ increase in ambient temperature. When assessing the influence of temperature on reaction rate, it is convenient to compute a quantity known as the **temperature quotient** ($Q_{10}$). This is arbitrarily defined as the ratio of the rate (R) of a process caused by a $10°C$ increase in temperature. For temperature intervals of $10°C$, we compute the $Q_{10}$ by dividing the reaction rate at the higher temperature (t+10) by reaction rate at the lower temperature (t):

$$Q_{10} = \frac{R_{(t+10)}}{R_{(t)}}.$$

In living systems, the relationship we have described between temperature and the rates of physiological reactions only holds true within certain limits. As you probably know, living organisms generally cannot tolerate ambient temperatures greater than about $50°C$. Reactions that are catalyzed by organic catalysts (**enzymes**) (see Exercise 2), show a marked decline in reaction rate when the reactants and catalysts are heated to high temperatures, because the protein structures of enzymes are irreversibly altered if they are heated beyond a certain point. This change in protein structure is called **denaturation** and is one reason why animals are unable to tolerate high ambient temperatures.

## Procedure

Record your results in the laboratory report section at the end of this exercise.

### Part 1. Effect of temperature on the ventilation rate of a frog

1. Obtain a live frog and a small glass jar. Partially fill the jar with tap water at about $20°C$. Use enough water to allow the frog to rest comfortably with its head above the water.

2. Place the frog in the jar and cover with a perforated lid. Put the jar in a culture dish and fill the culture dish with water at room temperature (about $20°C$) to the level of the water in the jar (Fig. 1-1).

3. Place a thermometer through one of the holes in the lid so that the thermometer bulb is in the water. Wait about 3 min and then record the temperature of the water.

4. Count the number of times the floor of the frog's mouth is lowered in 1 min. Record your data in Table 1-1 .

5. Repeat Step 4 two more times. Record your observations in Table 1-1.

6 Calculate the average value for ventilation rate at this temperature and record this value.

7. Add crushed ice to the culture dish. Note the temperature of the water in the jar after three minutes. If necessary, add more crushed ice to bring the temperature of the water in the jar to $15°C$. Maintain this temperature (plus or minus $1°C$, that is, $14°C$-$16°C$) by adding ice as the temperature of the solution rises.

8. Five or 6 min after the temperature has reached 15°C, measure the temperature and the ventilation rate. Repeat this procedure two more times. Enter your results in Table 1-1.

9. Repeat the procedure you used in Step 7 to lower the temperature of the water in the jar to 10°C. <u>Be sure to allow at least 5-6 min for temperature equilibration each time the temperature is changed.</u> Again, measure the ventilation rate three times and record your data.

10. Bring the temperature in the jar to 5°C and make another set of readings.

11. Enter all results in Table 1-1 and calculate average values for ventilation rate at each temperature.

12. Fill a 1-l beaker with water at 25°C. (The easiest way to do this is by mixing warm tap water and cold water until you get the desired temperature.) Pour off the water in the culture dish and pur about 2 cm of the warmed water into the culture dish. As before, maintain the temperature of the water in the jar within a degree of the desired temperature. Monitor the increase in temperature in the jar and allow 5-6 min for equilibration after the temperature reaches 25°C.

13. Record the ventilation rate three times, as you did in Step 9. Enter your values in Table 1-1.

14. Repeat Steps 12 to 14 using water heated to 30°C. (If water from the tap is not hot enough, it may be necessary to use a heated water bath or to heat a beaker of water over a Bunsen burner.)

15. Calculate average values for ventilation rate at each temperature.

Figure 1-1. *Apparatus for determining the effect of ambient temperature on the ventilation rate of a frog.*

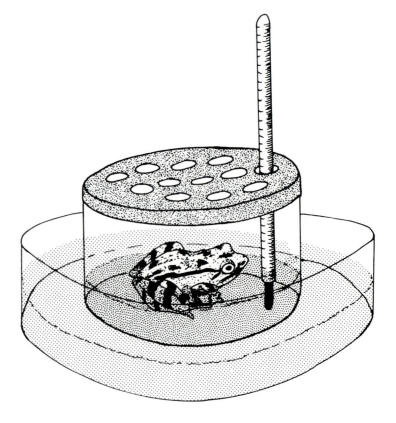

*Part 2. Effect of temperature on frog heart rate*

1. Your instructor will <u>double pith</u> a frog for each group (Fig. 1-2). Pithing rapidly destroys the brain so that the animal will feel no pain during the following experiment.

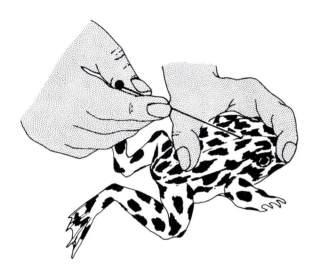

*Figure 1-2. Procedure for pithing a frog. Probe is inserted through the foramen magnum of the skull. This procedure is known as the **single pith**. It destroys the frog's brain while leaving the spinal cord intact so that spinal reflexes are not interfered with. To perform a **double pith**, the position of the probe is then reversed, and it is inserted into the spinal cord.*

2. Pin the frog to the frog board with the ventral side up.

3. Make an incision in the abdominal skin along the midline.

4. Cut two incisions at the anterior end of this slit, going laterally at right angles from the medial slit (Fig. 1-3A).

5. Repeat Step 4 at the posterior end of the medial incision.

6. Turn back the skin on both sides and, beginning at the posterior end of the exposed area, carefully make a cut through the abdominal muscles, a little to one side of the midline, to avoid cutting through the abdominal vein. Continue cutting until you reach the pectoral girdle.

7. Remove the sternum, as follows: Gently lift the sternum with your forceps. Cut through the bones at the base of each arm. Lift the sternum and cut its attachment at the throat. This will expose the beating heart (Fig. 1-3B).

8. Carefully remove the heart and transfer it to a small glass dish filled with Ringer's solution.

9. Wait until the heart resumes beating. If the heart does not begin to beat when you first put it in the culture dish, wait a few minutes. If it does not begin beating within a few minutes, you may have accidentally severed the thin-walled atria containing the heart's pacemaker. If the heart does begin beating, wait at least 10 min from the time you placed the heart in the dish.

10. Immerse a thermometer in the dish of Ringer's and determine the temperature (usually about $20°C$).

11. Measure the heart rate by counting the number of beats per min. Repeat this measurement twice. Enter your values in Table 1-2.

12. Calculate the average value for heart rate at this temperature and enter this value in Table 1-2.

13. Use small chips of ice made from Ringer's solution to lower the temperature of the solution bathing the heart to $15°C$. As before, maintain this temperature (plus or minus $1°C$, that is, $14°C-16°C$) by adding small Ringer's ice chips as necessary. Use the thermometer to monitor the temperature and to stir the solution during this operation.

14. After 5-6 min measure the temperature and the heart rate as above twice. Be sure to allow

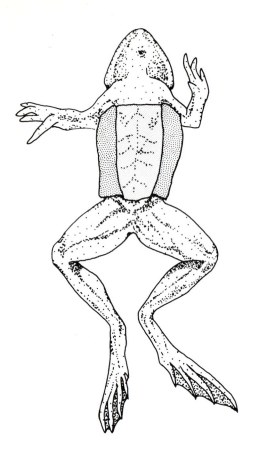

*Figure 1-3A.* Dissection of a frog to expose the abdominal musculature.

*Figure 1-3B.* Dissection of a frog to reveal the heart.

enough time for temperature equilibration each time the temperature is changed.

15. Now lower the temperature to 10°C, allow the heart to equilibrate, and repeat your measurements.

16. Make another pair of readings at 5°C.

17. Enter all results in Table 1-2 and calculate average values for heart rate at each temperature.

18. Use a beaker of Ringer's solution heated to 50°C over a Bunsen burner to increase the temperature of the bathing solution to 5° above room temperature. As before, maintain the solution within a degree of the desired temperature.

19. Allow a full 10 min for equilibration and then record the heart rate twice, as you did in Part 1. Enter your values in Table 1-2.

20. If time permits, increase the temperature of the Ringer's solution to 30°C and then to 35°C and note the responses of the frog's heart under these conditions. Enter your values in Table 1-2.

 *Part 3. The effect of temperature on tadpole heart rate*

1. Place a tadpole of the clawed frog (*Xenopus*) in about 2 cm of water at 20°C in a fingerbowl. With the aid of a dissecting microscope, observe the heart pulsating.

2. Obtain three 1 min readings, as before, and enter your values in Table 1-3.

3. Using the same procedure you used in Part 2, lower the temperature of the water by adding ice chips and make successive determinations of heart rate at temperatures of 15°C, 10°C, and 5°C. Allow time for equilibration each time you change the ambient temperature. Obtain three readings and calculate average heart rate for each temperature. Enter your data in Table 1-3.

4. Using the same procedure you used in Part 2, raise the temperature of the water first to 25°C, and then to 30°C. Make three successive readings at each temperature, and then allow time for equilibration at the new temperature. Enter your data in Table 1-3 and calculate average values for each ambient temperature.

# Laboratory Report

## Exercise 1: Effect of Temperature on Chemical Reactions in Cells

*Name:*

*Date:*

*Lab Section:*

## Analyzing Your Data

Using your data from Table 1-1, prepare a graph in which ventilation rate (beats/min) is plotted as a function of ambient temperature (°C). You want this graph to show how temperature affects ventilation rate, so ventilation rate is your dependent variable, shown on the vertical ($y$) axis, and

***Table 1-1.*** *Effect of ambient temperature on ventilation rate in the frog.*

| | Ventilation rate (beats/min) | | | |
|---|---|---|---|---|
| Temperature (°C) | Trial 1 | Trial 2 | Trial 3 | Average |
| **Temperatures above room temperature:** | | | | |
| 30 | | | | |
| 25 | | | | |
| **Room temperature:** | | | | |
| 20 | | | | |
| **Temperatures below room temperature:** | | | | |
| 15 | | | | |
| 10 | | | | |
| 5 | | | | |

temperature is your independent variable, plotted on the horizontal ($x$) axis. (See Appendix 1.) Prepare two additional graphs in which you show the relationship between heart rate and temperature for a frog (Table 1-2) and a tadpole (Table 1-3).

Calculate $Q_{10}$ values as follows: compute the ratio of frog ventilation rate (VR) at 15°C ($VR_{15}$) to ventilation rate at 5°C ($VR_5$). Enter this value ($VR_{15}/VR_5$) in Table 1-4.

Similarly, calculate the ratio of ventilation rate at 25°C ($VR_{25}$) to ventilation rate at 15°C ($VR_{25}/VR_{15}$) and the ratio of ventilation rate at 35°C ($VR_{35}$) to ventilation rate at 25°C ($VR_{35}/VR_{25}$).

In the same manner, compute $HR_{15}/HR_5$, $HR_{25}/HR_{15}$, $HR_{35}/HR_{25}$ for the tadpole heart and the frog heart. Enter all these $Q_{10}$ values in Table 1-4.

**Table 1-2.** *Effect of ambient temperature on heart rate in the frog.*

| Temperature (°C) | Heart rate (beats/min) | | | |
| --- | --- | --- | --- | --- |
| | Trial 1 | Trial 2 | Trial 3 | Average |
| **Temperatures above room temperature:** | | | | |
| 35 | | | | |
| 30 | | | | |
| 25 | | | | |
| **Room temperature:** | | | | |
| 20 | | | | |
| **Temperatures below room temperature:** | | | | |
| 15 | | | | |
| 10 | | | | |
| 5 | | | | |

*Table 1-3.* *Effect of ambient temperature on heart rate in the tadpole.*

| Temperature (°C) | Heart rate (beats/min) | | | |
|---|---|---|---|---|
| | Trial 1 | Trial 2 | Trial 3 | Average |
| **Temperatures above room temperature:** | | | | |
| 30 | | | | |
| 25 | | | | |
| **Room temperature:** | | | | |
| 20 | | | | |
| **Temperatures below room temperature:** | | | | |
| 15 | | | | |
| 10 | | | | |
| 5 | | | | |

**Table 1-4.** *$Q_{10}$ values for heart rate (HR) and ventilation rate (VR).*

| Temperature differential (°C) | Numerical value of ratio |
| --- | --- |
| **Ventilation rate:** | |
| $VR_{15}/VR_5$ | |
| $VR_{25}/VR_{15}$ | |
| $VR_{35}/VR_{25}$ | |
| **Tadpole heart rate:** | |
| $HR_{15}/HR_5$ | |
| $HR_{25}/HR_{15}$ | |
| $HR_{35}/HR_{25}$ | |
| **Frog heart rate:** | |
| $HR_{15}/HR_5$ | |
| $HR_{25}/HR_{15}$ | |
| $HR_{35}/HR_{25}$ | |

# *Questions*

1. What is kinetic energy?

2. What is activation energy?

3. Why do the rates of chemical reactions increase as ambient temperature increases?

4. What is meant by the term **poikilothermic**?

5. Why is the term **cold-blooded** misleading?

6. What limits the range of temperatures poikilothermic animals can tolerate?

7. What is the relationship between ventilation rate in a frog and ambient temperature?

8. a) What is the relationship between heart rate in a tadpole and ambient temperature? Explain the reasons for this relationship.

b) What is the relationship between heart rate in a frog and ambient temperature? Explain the reasons for this relationship.

9. Is the slope of the curve depicting the relationship between heart rate in the frog and ambient temperature constant?

_____

10. a) How does $VR_{35}/VR_{25}$ compare to $VR_{25}/VR_{15}$ for a frog?

b) How does $VR_{25}/VR_{15}$ compare to $VR_{15}/VR_5$ for a frog?

11. a) Did the $Q_{10}$ values for heart rate in the tadpole show the same relationships you described in your answer to Question 10a? _____ Explain.

b) Did the $Q_{10}$ values for heart rate in the frog show the same relationships you described in your answer to Question 10b? _____ Explain.

12. What property of the frog heart caused it to continue beating after it was removed from the frog?

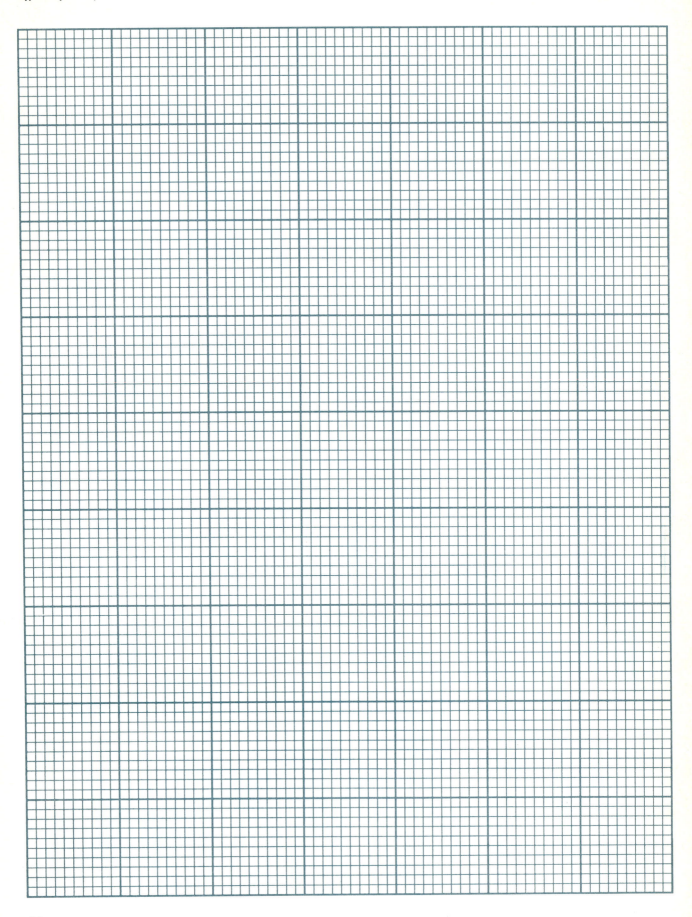

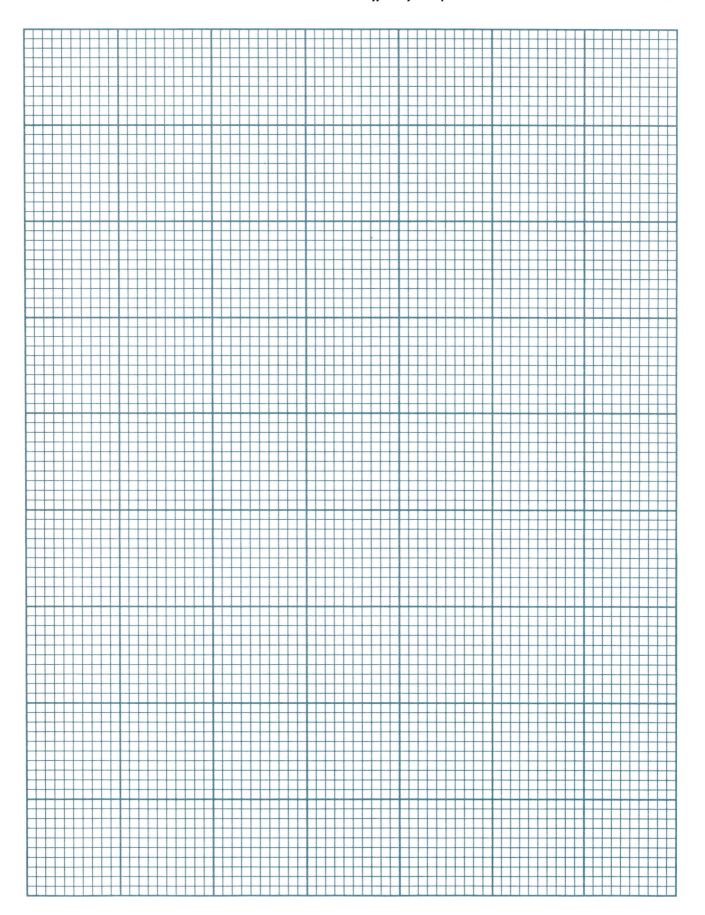

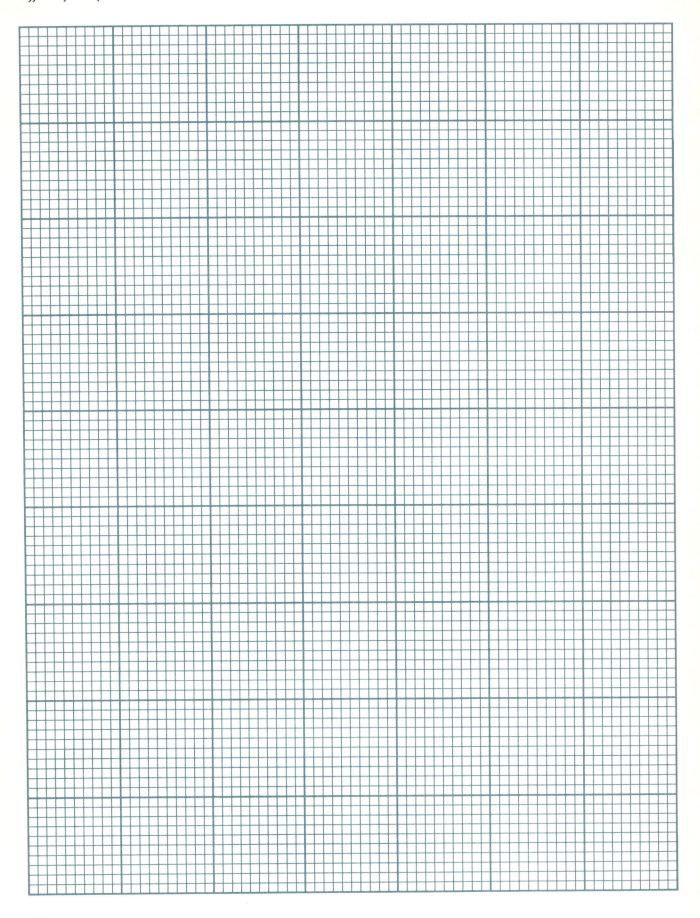

# Action of Enzymes

*Reading assignment: text 26-27, 35, 48, 50-53*

## Objectives

### Experimental

1. To observe the hydrolysis of nitrophenyl phosphate by alkaline phosphatase.

2. To determine the effect of temperature on the activity of alkaline phosphatase.

3. To determine the effect of pH on the activity of alkaline phosphatase.

### Conceptual

After completing this exercise and the reading assignment, you should be able to:

1. Define the terms **enzyme, free energy, activation energy, denaturation, substrate, product, hydrolysis, transmittance, absorbance,** and **buffer.**

2. List four characteristics of enzymes.

3. Describe the effect of temperature on enzyme action and explain why it has this effect.

4. Describe the effect of pH on enzyme action and explain why it has this effect.

5. Explain why absorbance at 405 nm can be used as an indicator of the activity of alkaline phosphatase.

6. Discuss the relationship between transmittance and absorbance.

7. Explain what a standard curve is, how it is determined, and how it can be used.

8. Discuss the physiological importance of enzymes.

## Background

**Enzymes** are organic catalysts, proteins that increase the rates of chemical reactions. The portion of a system's total energy that is available to do work is its **free energy**. Chemical reactions involve **transition states**, whose free energy is higher than the total free energy of either the reactants or the products (Fig. 2-1A). The difference between the free energy of the reactants and the free energy of the transition state is called the **activation energy** of a reaction. Enzymes reduce the **activation energy** of the reactions they catalyze

(Fig. 2-1B). Because they are neither consumed nor modified in the chemical reactions they affect (Fig. 2-2), small quantities of enzyme can mediate the formation of large quantities of product.

Enzymes are proteins. The three-dimensional structure of proteins can be altered by changes in temperature or pH (Fig. 2-3). These changes in three-dimensional structure are called **denaturation**. In Parts 3 and 4 of this exercise you will explore the effects of temperature and pH on enzyme action.

*Action of Enzymes*

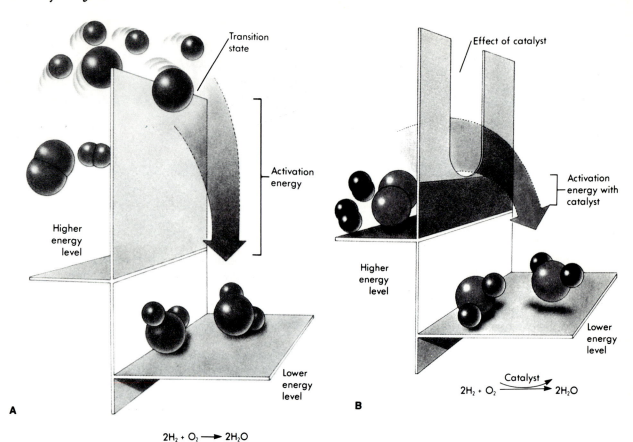

$2H_2 + O_2 \longrightarrow 2H_2O$

$2H_2 + O_2 \xrightleftharpoons{\text{Catalyst}} 2H_2O$

***Figure 2-1. A*** *Chemical reactions involve intermediate transitional states that can be viewed as energy barriers. The height of the barrier is the activation energy.* ***B*** *Enzymes catalyze chemical reactions by reducing the activation energy.*

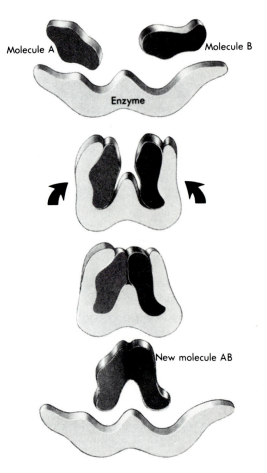

***Figure 2-2.*** *Enzymes have active sites that bind substrates. The active sites must recognize some aspect of the molecular structure of their substrates. While they participate in the reaction, enzymes are not permanently altered and can catalyze further reactions.*

20

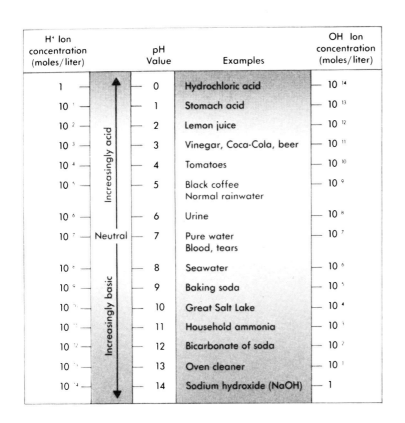

| H$^+$ Ion concentration (moles/liter) | | pH Value | Examples | OH Ion concentration (moles/liter) |
|---|---|---|---|---|
| 1 | ↑ Increasingly acid | 0 | Hydrochloric acid | 10$^{14}$ |
| 10$^1$ | | 1 | Stomach acid | 10$^{13}$ |
| 10$^2$ | | 2 | Lemon juice | 10$^{12}$ |
| 10$^3$ | | 3 | Vinegar, Coca-Cola, beer | 10$^{11}$ |
| 10$^4$ | | 4 | Tomatoes | 10$^{10}$ |
| 10$^5$ | | 5 | Black coffee / Normal rainwater | 10$^9$ |
| 10$^6$ | | 6 | Urine | 10$^8$ |
| 10$^7$ | Neutral | 7 | Pure water / Blood, tears | 10$^7$ |
| 10$^8$ | ↓ Increasingly basic | 8 | Seawater | 10$^6$ |
| 10$^9$ | | 9 | Baking soda | 10$^5$ |
| 10$^{10}$ | | 10 | Great Salt Lake | 10$^4$ |
| 10$^{11}$ | | 11 | Household ammonia | 10$^3$ |
| 10$^{12}$ | | 12 | Bicarbonate of soda | 10$^2$ |
| 10$^{13}$ | | 13 | Oven cleaner | 10$^1$ |
| 10$^{14}$ | | 14 | Sodium hydroxide (NaOH) | 1 |

*Figure 2-3. The pH scale. At pH 7.0, the concentrations of H$^+$ and OH$^-$ are equal; a solution with a pH of 7.0 is said to be neutral. Solutions with pH below 7.0 are acidic; solutions with pH greater than 7.0 are basic, or alkaline. Blood and most other bodily fluids are slightly alkaline.*

Without enzymes, most physiological processes would proceed at extremely slow rates. Enzymes are important in the maintenance of cellular resting potentials, the contraction of muscles, and in all stages of metabolism, from the breakdown, digestion, and assimilation of food to cellular metabolism and the excretion of waste products.

The enzyme we will work with in this exercise is normally produced by the membranes of cells lining the small intestine. It is a **phosphatase**, an enzyme that catalyzes the removal of phosphate groups (PO$_4^{-2}$) from some organic molecules. A number of biologically important compounds, including DNA, contain phosphates. If we let R represent the rest of the molecule, phosphates can be designated

R-PO$_4$ .
phosphate group

Because this enzyme works best under the non-acid conditions found in the small intestine, its common name is **alkaline phosphatase**.

When phosphates react with water, the water molecule splits the phosphate as follows:

$$\text{R-PO}_4 + \text{H}_2\text{O} \xrightarrow{\text{alkaline phosphatase}} \text{ROH} + \text{H}_3\text{PO}_4 .$$
SUBSTRATE · · · · · · · · · · · · · · · PRODUCT

In this reaction the phosphate (R-PO$_4$) is the **substrate** upon which the enzyme, alkaline phosphatase, acts, and ROH and H$_3$PO$_4$ are the **products**. The process of splitting a molecule by the addition of a molecule of water is known as **hydrolysis**.

If we have some way to measure the amount of ROH, we can follow the progress of the reaction. Fortunately, there is an organic phosphate that has the characteristics we need. **Nitrophenyl phosphate** is colorless (Fig. 2-4A). However,

**\*\* when the substrate nitrophenyl phosphate is hydrolyzed by the enzyme, alkaline phosphatase (Fig. 2-4B), the product (ROH, or nitrophenol) is yellow:**

$$\text{nitrophenyl phosphate} + H_2O \xrightarrow{\text{alkaline phosphatase}} \text{nitrophenol} + H_3PO_4.$$

nitrophenyl phosphate
COLORLESS

nitrophenol
YELLOW

So the progress of the reaction is signalled by an increasingly yellow color in the originally colorless solution. The faster the color develops, the faster the reaction must be running. This means that the progress of the enzyme-catalyzed reaction can be followed by monitoring the change from colorless to yellow in the test solutions. One way to do this is by visual inspection. You will notice the development of a yellow color in those tubes in which nitrophenyl phosphate is hydrolyzed to nitrophenol.

More information about the amount of nitrophenol produced can be obtained if the intensity of the yellow color produced is measured with a **spectrophotometer,** or colorimeter (Fig. 2-5). The spectrophotometer separates white light into its component colors by means of a prism or a diffraction grating. When you use this instrument, you

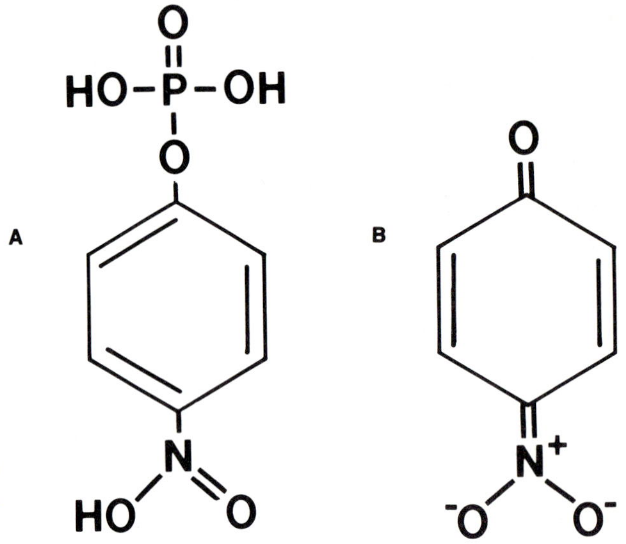

*Figure 2-4. Structure of p-nitrophenyl phosphate (A) and p-nitrophenol (B).*

select light of a specified wavelength and shine it through a solution contained in a special tube called a **cuvette.**

When a solution absorbs light at some wavelengths and not others, we see the light that is <u>not</u> absorbed. Nitrophenol absorbs purple light (wavelength 405 nm); therefore, it appears yellow. The colorless compound nitrophenyl phosphate does not absorb light at this wavelength. Thus, the amount of light absorbed at 405 nm is an indica-

*Figure 2-5. A spectrophotometer.*

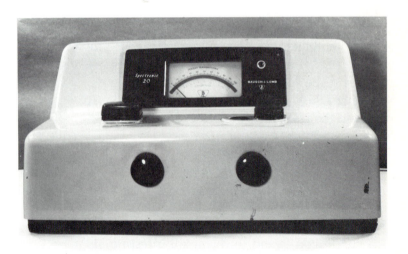

tion of the intensity of the yellow color, and hence of the amount of nitrophenol, that is produced.

A spectrophotometer enables you to determine the amount of light that passes through a solution (the **transmittance**), as well as its **absorbance**, a measure of the amount of light that is absorbed by the solution. Transmittance and absorbance are inversely related to each other. If most of the light passes through a solution, little light is absorbed. Such a solution has high transmittance and low absorbance. Similarly, if much light is absorbed, little light will be transmitted.

For many solutions absorbance is directly proportional to the concentration of pigment in the solution. You will take advantage of this fact in this exercise. In Part 1, the class will determine the absorbance of several solutions containing known concentrations of the product, nitrophenol. You will then use these values to plot absorbance as a function of the concentration of nitrophenol. This will give you a **standard curve** (Fig. 2-6) for nitrophenol. You will use this curve to determine the concentration of product in the test solutions from other parts of this exercise.

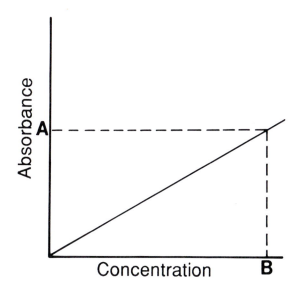

*Figure 2-6. Determination of concentration of pigment in an unknown solution using a standard curve. The concentration of a solution with absorbance equal to A is B.*

## Procedure

Record your results in the laboratory report section at the end of this exercise.

### *Part 1. Preparation of a standard curve for nitrophenol*

**1. Obtain two cuvettes. Oils from your fingers can alter the amount of light transmitted through the cuvette. Therefore, you should not handle the lower half of the tube, and you should wipe cuvettes with a lint free cloth or tissue before placing them in the spectrophotometer.**

**2. Make sure the lid of the sample holder in the spectrophotometer is closed. While there is no cuvette in the cuvette holder, turn the amplifier control knob to set the pointer at ∞ <u>absorbance</u>. (There will be two scales on the spectrophotometer. "A" is for absorbance; "T" is for transmittance. In this exercise you will need to refer only to the absorbance scale.)**

**3. Fill the cuvettes with the following (Fig. 2-7):**

> **Cuvette 1: 5 ml distilled water. This is your <u>blank</u>.**

> **Cuvette 2: 5 ml of one of the <u>standard</u> nitrophenol solutions.**

**The concentrations of the nitrophenol standards are:**

> **2 μg/ml,**
> **4 μg/ml,**
> **6 μg/ml,**
> **8 μg/ml,**
> **10 μg/ml.**

**4. Wipe the outside of the cuvette to dry it.**

**5. Place the cuvette containing the blank (Cuvette 1) in the sample holder. The cuvette will be marked with a line or a triangle. Make sure that this faces toward the front of the**

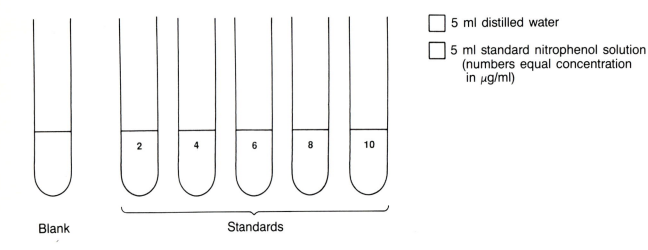

5 ml distilled water

5 ml standard nitrophenol solution (numbers equal concentration in μg/ml)

Blank      Standards

*Figure 2-7. Experimental set-up for determining standard curve for nitrophenol.*

spectrophotometer. (The cuvette holder may contain a line with which you can line up the mark on the cuvette.)

6. Close the lid of the sample holder. Using the light control knob, set the pointer at 0% absorbance.

9. Remove the blank from the sample holder.

10. Insert Cuvette 2 (the standard) into the sample holder, close the lid, and read the percent absorbance of the standard. Enter this value in Table 2-1 and on the board.

11. Other members of your class will determine the percent absorbance of other concentrations of nitrophenol, and these values will be written on the board. Enter these values in Table 2-1.

 *Part 2. The hydrolysis of nitrophenyl phosphate by alkaline phosphatase*

1. Using a wax pencil, number four test tubes from 1 to 4. For each tube, record the time when you combine the reagents, to the nearest 30 sec.

2. Add reagents to the four test tubes as follows (Fig. 2-8):

Tube 1: Substrate (nitrophenyl phosphate) + buffer + water. Pipette 2 ml of the substrate into Tube 1. Add 2 ml buffer (pH 10) and 2 ml water. Place in test tube rack.

Tube 2: Substrate (nitrophenyl phosphate) + buffer + enzyme. Pipette 2 ml substrate, 2 ml buffer (pH 10), and 2 ml enzyme into Tube 2. Place in rack.

Tube 3: Substrate (nitrophenyl phosphate) + buffer + boiled enzyme. Pipette

2 ml substrate, 2 ml buffer (pH 10), and 2 ml boiled enzyme into Tube 3. Place in rack.

Tube 4: <u>Reaction product (nitrophenol) + water</u>. Pipette 2 ml of the product, nitrophenol, plus 4 ml of water into Tube 4. The color in this tube will provide you with an <u>end point</u> (an example of how a solution will appear after all the substrate has been converted to the product).

3. Observe the tubes. If color develops in any of them, compare it from time to time with Tube 4, containing a known amount of reaction product. If the color of the solution in any of the tubes changes to match the nitrophenol solution (Tube 4), record the time at which the color change occurred in Table 2-2. After 10 min have elapsed, record the color in each tube in Table 2-2. Terminate your experiment after 10 min.

4. Rinse a cuvette with distilled water. Pour the substrate-buffer-water mixture from Tube 1 into this cuvette. This is your blank for the rest of this exercise.

5. Close the lid of the sample holder and set the pointer at 0% absorbance.

6. Pour the substrate-buffer-enzyme mixture from Tube 2 into a clean cuvette and wipe it dry.
7. Place this cuvette into the cuvette holder, close the lid, and read the percent absorbance for this

test solution from the "A" scale. Enter this value in Table 2-3.

8. Repeat Steps 4 and 5 with the substrate-buffer-boiled enzyme mixture in Tube 3. Enter the result in Table 2-3.

9. Repeat Steps 4 and 5 with the nitrophenol solution in Tube 4. Enter the result in Table 2-3.

 *Part 3. Effect of temperature on enzyme action*

1. Set up 4 test tubes marked 1-4. To each tube add 2 ml substrate and 2 ml buffer (pH 10).

2. Place test tubes in water baths of varying temperatures as follows (Fig. 2-9):

Tube 1: water bath cooled with ice to about $10^{\circ}$C.

Tube 2: water bath at room temperature ($20^{\circ}$C).

Tube 3: water bath heated to $37^{\circ}$C.

Tube 4: water bath heated to $50^{\circ}$C.

3. Add 2 ml enzyme to Tube 1 and determine the time it takes for the color to equal that in Tube 4, Part 1.

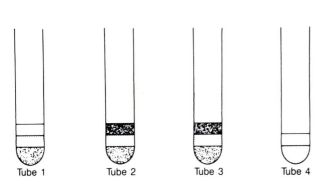

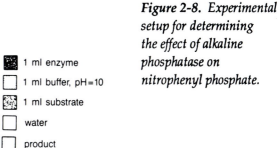

*Figure 2-8. Experimental setup for determining the effect of alkaline phosphatase on nitrophenyl phosphate.*

## Action of Enzymes

*Figure 2-9. Experimental setup for determining effect of temperature on activity of alkaline phosphatase.*

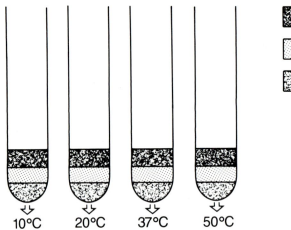

10°C   20°C   37°C   50°C

■ 1 ml enzyme

□ 1 ml buffer, pH=10

▨ 1 ml substrate

4. Repeat for tubes 2-4. Record your results in Table 2-4.

5. Measure the absorbance at 405 nm of the solutions in tubes 1-4, using the procedure described in steps 4 and 5 of Part 1. Enter your results in Table 2-5.

 **Part 4. Effect of pH on enzyme action**

1. All tubes in this part of the exercise are to be kept at room temperature. Set up 5 test tubes coded 1-5.

2. To each tube add 2 ml of substrate (Fig. 2-10).

3. Add buffers as follows:

Tube 1: 2 ml pH 7 buffer.

Tube 2: 2 ml pH 8 buffer.

Tube 3: 2 ml pH 9 buffer.

Tube 4: 2 ml pH 10 buffer.

Tube 5: 2 ml pH 11 buffer.

4. Add 2 ml enzyme to each tube and determine the time for color development. Enter your results in Table 2-6.

5. Measure the absorbance at 405 nm of the solutions in tubes 1-5, using the procedure described in steps 4 and 5 of Part 1. Enter your results in Table 2-7.

*Figure 2-10. Experimental setup for determining effect of pH on activity of alkaline phosphatase.*

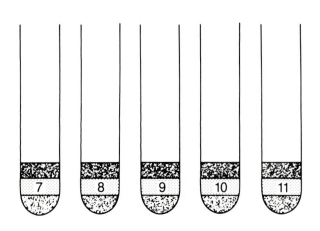

7   8   9   10   11

■ 1 ml enzyme

□ 1 ml substrate

▨ 1 ml buffer (number=pH)

# Laboratory Report

## Exercise 2: Action of Enzymes

Name: _____

Date: _____

Lab Section: _____

## Analyzing Your Data

Inspection of your data in Table 2-2 should tell you whether your experiment showed that alkaline phosphatase catalyzes the hydrolysis of nitro-phenyl phosphate to nitrophenol.

To calculate reaction rate from your data, deter-mine the amount of nitrophenol appearing per minute, as follows:  The reference tube (Tube 4, Part 2) contains 10 micrograms (µg) of nitrophe-nol.  Each experimental tube in which there was a color change to the same color intensity as the end

point must also have contained 10 µg nitrophenol. The rate at which this was produced must be:

$$\frac{10\ \mu g}{\text{number of min to produce color}} = \frac{X\ \mu g}{\text{min}}$$

Using your data from tables 2-4 and 2-6 and the graph paper provided at the end of this exercise, plot reaction rate as a function of 1) temperature and 2) pH.  (Make two separate graphs.)

Next, use the data from Table 2-1 to plot absorbance at 405 nm as a function of the concentration of

Table 2-1. *Absorbance of different concentrations of nitrophenol.*

| Concentration of nitrophenol (µg/ml) | Absorbance |
|---|---|
| 2 | |
| 4 | |
| 6 | |
| 8 | |
| 10 | |

nitrophenol. This is your standard curve. It will enable you to convert the values for absorbance recorded in Tables 2-3, 2-5, and 2-7 to concentrations. Do this, and enter these values in the appropriate tables.

Now you have the information you need to plot the concentration of the reaction product as a function of temperature (Table 2-5) and as a function of pH (Table 2-7). The concentration of product is a measure of enzyme activity.

*Table 2-2. The effect of alkaline phosphatase on nitrophenyl phosphate.*

| Tube | Treatment | Starting time (min) | Ending time (min) | Time elapsed (min) | Color in tube after 10 min |
|------|-----------|---------------------|-------------------|--------------------|-----------------------------|
| 1 | Substrate + buffer + $H_2O$ | | | | |
| 2 | Substrate + buffer + enzyme | | | | |
| 3 | Substrate + buffer + boiled enzyme | | | | |

*Table 2-3. Absorbance at 405 nm of nitrophenyl phosphate (substrate) combined with buffer, alkaline phosphatase, and boiled alkaline phosphatase.*

| Treatment | Absorbance | Concentration of product ($\mu$g/ml)[1] |
|-----------|------------|------------------------------------------|
| Substrate + buffer + $H_2O$ | | |
| Substrate + buffer + enzyme | | |
| Substrate + buffer + boiled enzyme | | |

[1] Determined from standard curve.

***Table 2-4.*** *The effect of temperature on enzyme action.*

| Tube | Temperature | Starting time (min) | Ending time (min) | Time elapsed (min) | Color in tube after 10 min |
|------|-------------|---------------------|-------------------|--------------------|----------------------------|
| 1 | $10^\circ$C | | | | |
| 2 | $20^\circ$C | | | | |
| 3 | $37^\circ$C | | | | |
| 4 | $50^\circ$C | | | | |

***Table 2-5.*** *The effect of temperature on absorbance of test solutions containing alkaline phosphatase and nitrophenyl phosphate.*

| Temperature | Absorbance | Concentration of product ($\mu$g/ml)[1] |
|-------------|------------|------------------------------------------|
| $10^\circ$C | | |
| $20^\circ$C | | |
| $37^\circ$C | | |
| $50^\circ$C | | |

[1] Determined from standard curve.

*Table 2-6. The effect of pH on enzyme action.*

| Tube | pH of Buffer | Starting time (min) | Ending time (min) | Time elapsed (min) | Color in tube after 10 min |
|------|--------------|---------------------|-------------------|--------------------|----------------------------|
| 1 | 7 | | | | |
| 2 | 8 | | | | |
| 3 | 9 | | | | |
| 4 | 10 | | | | |
| 5 | 11 | | | | |

*Table 2-7. The effect of pH on absorbance of test solutions containing alkaline phosphatase and nitrophenyl phosphate.*

| pH | Absorbance | Concentration of product ($\mu$g/ml)[1] |
|----|------------|-----------------------------------------|
| 7 | | |
| 8 | | |
| 9 | | |
| 10 | | |
| 11 | | |

[1] Determined from standard curve.

# Questions

1. What are enzymes?

2. How do enzymes produce their effects?

3. What color is light with a wavelength of 405 nm?

   _____

4. What color are solutions that absorbs at 405 nm?

   _____

5. Why is absorbance at 405 nm an indication of the activity of alkaline phosphatase?

6. Choose the correct answer: If a solution has high transmittance at a given wavelength, its absorbance will be (high, low).

   _____ Explain.

7. Why did the blank used in parts 2 to 4 contain substrate, buffer, and water?

8. How do you know that the color changes you observed were due to alkaline phosphatase? (Hint: what were the controls in this experiment?)

9. a) Estimate the optimum temperature for alkaline phosphatase from your results. _____

   b) Was the optimum temperature for alkaline phosphatase what you expected? _____ Explain.

10. On the basis of your results, estimate the optimum pH for alkaline phosphatase. _____

11. Was the optimum temperature for alkaline phosphatase what you expected? (Hint: The pH of intestinal contents is 7.4 to 7.6.)

   _____

12. Which of the experimental treatments you used probably caused denaturation of the alkaline phosphatase?

13. Discuss the physiological importance of enzymes.

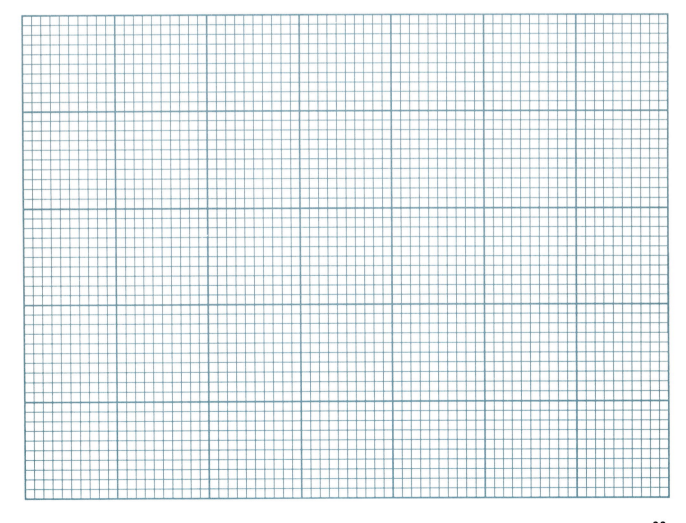

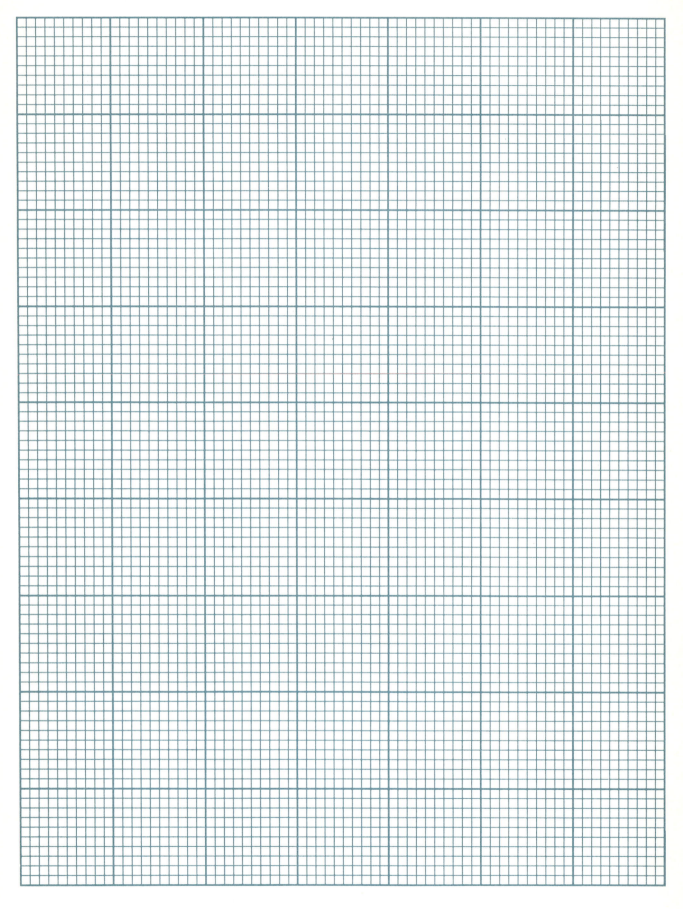

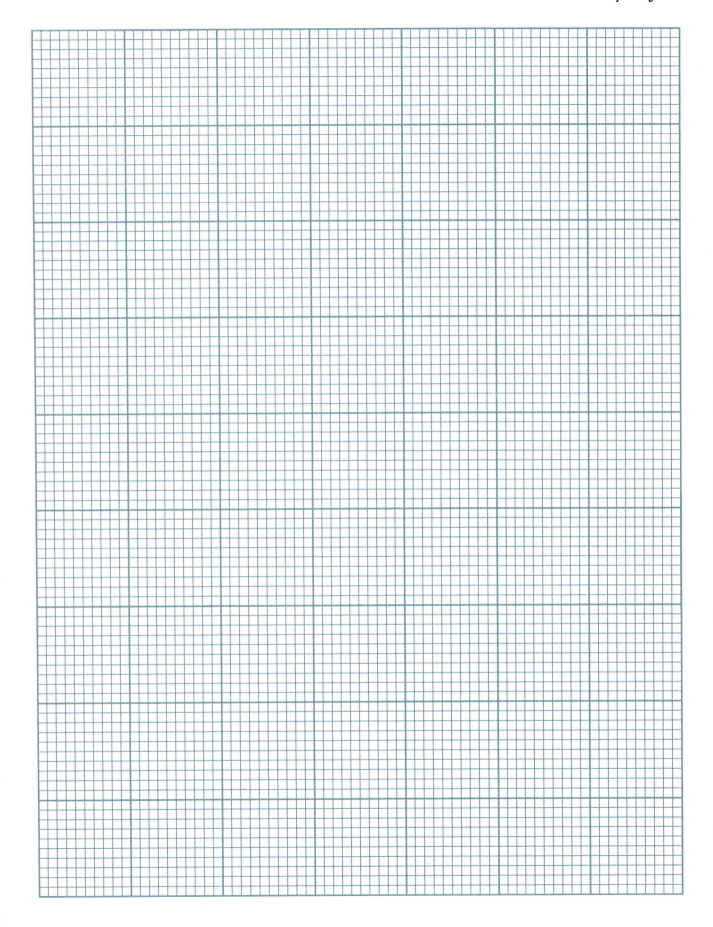

*Action of Enzymes*

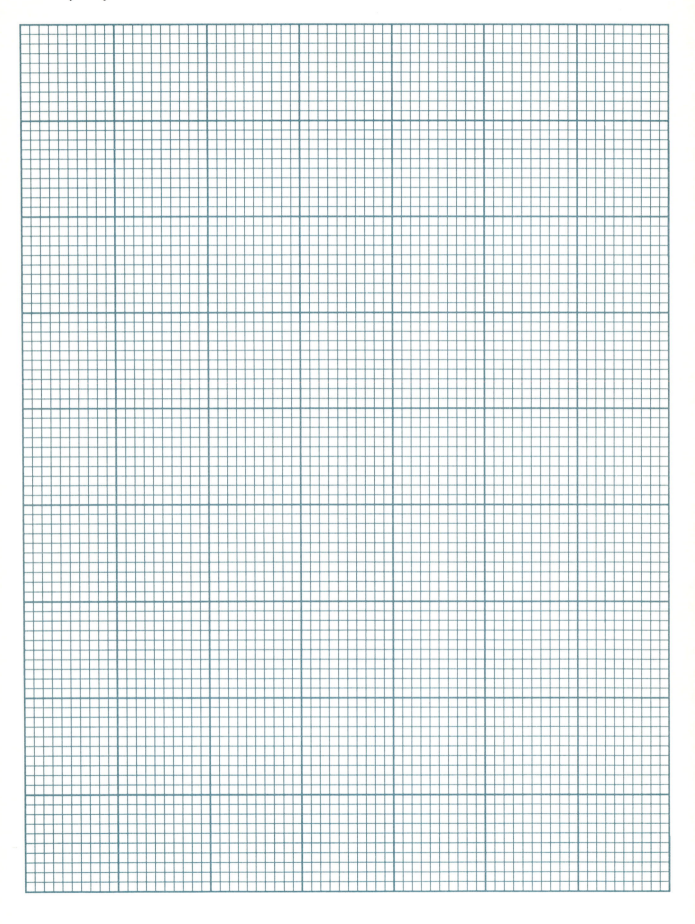

# Aerobic Metabolism: The Oxidation of Glucose

*Reading assignment: text 75-82*

## Objectives

 ### Experimental

1. To measure oxygen consumption in yeast for different concentrations of a) yeast and b) glucose.

2. To determine the relationship between the initial velocity of the reactions of aerobic metabolism in yeast and the concentration of a) yeast and b) glucose.

3. To determine the relationship between the final velocity of the reactions of aerobic metabolism in yeast and the concentration of a) yeast and b) glucose.

 ### Conceptual

After completing this exercise and the reading assignment, you should be able to:

1. Explain how changes in gas volume can be used to measure respiration rate.

2. Define the terms **substrate-level phosphorylation, oxidative metabolism, mitochondrion, crista, aerobic metabolism, active site, saturation, endogenous respiration**, and **exogenous respiration**.

3. Explain what a **manometer** is and how it can be used to measure cellular respiration rates.

4. Explain how the amount of respiration due to exogenous respiration can be calculated.

5. Describe and give one reason for the relationship between reaction velocity and yeast (enzyme) concentration.

6. Describe and give one reason for the relationship between reaction velocity and glucose (substrate) concentration.

## Background

Energy is required for many physiological processes, including active transport of substances across cell membranes, muscular contraction, and biochemical synthesis. The oxidation of glucose to carbon dioxide and water is a major source of energy. If glucose were broken down to carbon dioxide and water all at once, most of the energy released would be lost. Instead, the glucose in cells is broken down in a series of steps. The energy contained in the chemical bonds of glucose molecules is eventually transferred to phosphorylated nucleotides, such as adenosine triphosphate (ATP). The hydrolysis of ATP to adenosine diphosphate (ADP) and inorganic phosphate releases a convenient amount of energy for cells to use. There are two mechanisms for production of ATP. In **substrate-level phosphorylation**, a phosphate group is transferred to ADP from an intermediate compound to form ATP. The second mechanism of ATP production is **oxidative phos-**

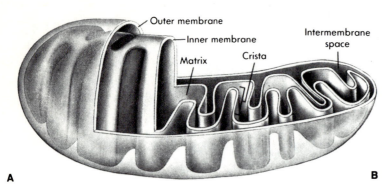

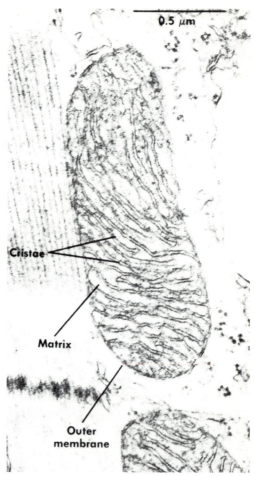

*Figure 3-1.* *A schematic illustration (A) and electron micrograph (B) of a mitochondrion. The enzymes of oxidative phosphorylation are located on the cristae, or inner membrane.*

**phorylation**. This process occurs in the inner membranes (**cristae**) of **mitochondria** (Fig. 3-1) and, as the term suggests, requires oxygen. Most of the ATP used by the body is produced by oxidative metabolism. Oxidative phosphorylation and its associated reactions are collectively called **aerobic metabolism**.

The biochemical pathways of oxidative metabolism are essentially the same in all organisms with nucleated cells. This includes animals, plants, fungi, and unicellular organisms except bacteria and blue-green algae. Because of this similarity we can use organisms to which we are only distantly related in the study of energy metabolism.

In this exercise, you will observe aerobic metabolism in yeast, a fungus. Yeast cells are used because they are convenient and reliable. Thin slices of animal tissue could be used, but it is difficult to be sure that reactants and oxygen are adequately delivered to pieces of tissue in which the blood supply has been disrupted.

Oxidative metabolism is the net effect of a number of enzymes, including those of glycolysis and oxidative metabolism and some that facilitate the diffusion of glucose into cells. Nevertheless, this experimental system approximates the behavior of a single enzyme fairly closely. In this exercise

you will vary the concentration of the enzymes of aerobic metabolism by varying the concentration of yeast.

As you know from Exercise 2, enzymes are proteins that catalyze specific chemical reactions. For enzymes to increase the rates of chemical reactions, molecules of the reactants must bind to a specific part of the enzyme called the **active site** (see Fig. 2-2). The reactants are converted to one or more products and released from the active site.

The rate of product formation is directly proportional to the number of enzyme-reactant complexes (Fig. 3-2). When the concentration of reactants is relatively low, the rate of an enzyme-catalyzed reaction depends on the concentration of reactants, the concentration of enzyme, and temperature. If enzyme concentration and temperature are held constant as the concentration of reactant is increased, the number of free active sites decreases. When all of the active sites are occupied by molecules of the reactants, no further increase in the rate of product formation can be achieved by further increases in concentration of the reactants. Under these conditions, the enzyme is said to be **saturated**. On the other hand, if you increase the concentration of yeast without changing the concentration of substrate, you increase the number of available active sites.

Many techniques are used to measure rates of cellular metabolism. A **manometer** is an instrument that measures changes in gas pressure (Fig. 3-3). When a gas is either consumed or released, the reaction can be run in a closed vessel, and the pressure changes resulting from gas production or consumption can be measured with a manometer. In the complete oxidation of a molecule of glucose, a molecule of one gas ($O_2$) is consumed, while a molecule of another gas ($CO_2$) is released:

$$C_6H_{12}O_6 + 6O_2 \rightarrow 6CO_2 + 6H_2O.$$
glucose

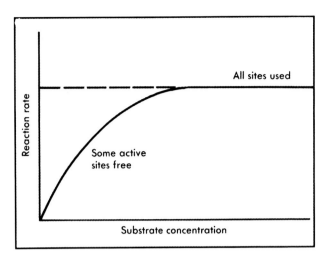

*Figure 3-2. Relationship between the reaction rate and the substrate concentration for a given concentration of enzyme.*

The gas removed is replaced by the gas produced. Therefore, except for changes in gas pressure caused by differences in the solubilities of $O_2$ and $CO_2$, there are no pressure changes to measure. However, if we convert the gaseous end-product to a water-soluble compound, then the sole change

*Figure 3-3. Manometer apparatus.*

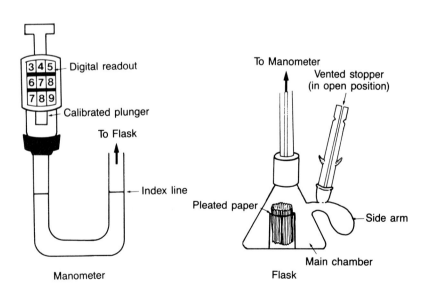

in the gas phase will be a decrease in $O_2$, and hence a decrease in gas pressure, as the reaction progresses. This can happen if a strong base, such as 10% potassium hydroxide (KOH) is included in the system. In that case, $CO_2$ will be converted to carbonate ion ($CO_3^{-2}$) and will remain in solution in this form.

In this exercise, the respiring yeast cells are enclosed in a gas-tight vessel attached to a manometer. Oxygen uptake by the cells will result in a decrease in the volume of gas in the vessel. This will cause a drop in pressure that you will measure with the manometer. The volume of gas consumed is measured by advancing a calibrated screw. Advancing the screw replaces the lost volume until the pressure is restored to its original value.

You will measure the effect on $O_2$ consumption of varying the concentration of a) substrate (glucose) for a fixed amount of yeast (Part 1) and b) enzyme (yeast) concentration for a fixed amount of glucose (Part 2). You will use your data to calculate initial reaction velocities and final reaction velocities for your experiments. This will enable you to assess the effects of substrate concentration and enzyme concentration on the rate of aerobic metabolism.

Respiration that uses substrate stored in the yeast cells is referred to as **endogenous respiration**. **Exogenous respiration** is the rate of respiration due to additional substrate that you add. If you know $O_2$ production due to endogenous respiration changes over time and total $O_2$ consumption over time you can calculate exogenous respiration by subtraction (Fig. 3-4). You will use this technique to determine how much oxygen consumption is stimulated by the exogenous glucose you add to the reaction vessel.

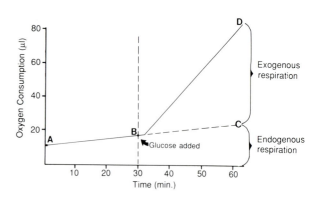

*Figure 3-4. Graph of oxygen consumption as a function of time in yeast. AB represents measured endogenous respiration. BC represents estimated endogenous respiration arrived at by extrapolation of line AB. BD represents total respiration after addition of glucose. Exogenous respiration is calculated by subtracting values on line BC from corresponding values on line BD.*

## Procedure

Record your results in the laboratory report section at the end of this exercise.

Eleven experiments will be necessary to provide all the data you will need to investigate the effects of yeast (enzyme) concentration (Part 1) and glucose (substrate) concentration (Part 2) on oxygen consumption. In Part 3 you will do an additional determination with a different concentration of either yeast or glucose. Since you cannot complete 11 experiments in one lab period, you will work in groups. Each group will do three determinations (combinations of yeast and glucose concentrations). There will be some replication. Your lab instructor will assign the determinations to each group.

While one person is taking readings, the other can be making up and equilibrating the next sample.

 ***Part 1. Effect of glucose concentration on oxygen consumption***

1. Familiarize yourself with the reaction chamber (Fig. 3-4). The center well will hold KOH. A piece of filter paper in the well will draw the KOH up by capillary action. The paper is fluted; the folds provide a large surface area for saturation with KOH. <u>Be sure that the KOH does not spill over into the main reaction chamber and that the filter paper does not touch the side of the flask.</u> The main chamber is for the yeast, and the side arm will hold the glucose.

2. Shake the <u>4 mg/ml yeast</u> suspension to ensure uniformity. Pipette 1.5 ml of yeast solution into the main chamber.

3. Put 0.1 ml of 10% KOH into the center well.

4. Put a pieced of pleated filter paper into the center well.

5. Pipette 1.0 ml of one of the following glucose solutions into the side arm:

> 10 µM/ml,
> 20 µM/ml,
> 30 µM/ml,
> 50 µM/ml,
> 75 µM/ml,
> 100 µM/ml.

6. Lightly grease the side-arm stopper and seal the side arm.

7. Couple the flask to the manometer.

8. Lower the flask into the water bath, turn on the shaker, and allow it to equilibrate for 10 min. During this time, the manometer should be open (shunted).

9. Close the manometer valve and take three successive readings at 5-min intervals. These give you a measure of endogenous respiration. Enter your data in Table 3-1.

10. Being careful of the KOH in the center well, tip the chamber so that only the glucose and the yeast are mixed. Your instructor will demonstrate this procedure. <u>This should be done very carefully.</u> Errors can easily be introduced at this time from movement of the flask. In addition, be sure that the connecting hose does not become crimped. Exogenous respiration starts as soon as the glucose and yeast come in contact with each other.

11. Tip the manometer again to pour material from the main chamber into the side arm.

12. Then tip it one more time to pour most of the reaction mixture back into the main chamber.

13. Read the volume of oxygen consumed at eight successive 3-min intervals. Enter your results in Table 3-1.

**Part 2. Effect of yeast concentration on oxygen consumption**

**1. Put 1 ml of one of the following yeast solutions (as assigned by your instructor) into the main chamber:**

> **1 mg/ml,**
> **2 mg/ml,**
> **4 mg/ml,**
> **6 mg/ml,**
> **8 mg/ml.**

**2. Put 0.1 ml of 10% KOH into center well and add filter paper.**

**3. Put 1.0 ml of 50 μM/ml glucose solution into the side arm.**

**4. Measure endogenous respiration and total respiration by repeating Part 1, steps 7 to 13. Enter your results in Table 3-2.**

**Part 3.**

**You have now done two determinations. Repeat either Part 1 or Part 2, depending upon whether the third determination assigned to your group involves varying the concentration of yeast or of glucose. Enter your results in Table 3-3.**

# Laboratory Report

*Exercise 3: Aerobic Metabolism:*
*The Oxidation of Glucose*

Name: _____

Date: _____

Lab Section: _____

## Analyzing Your Data

The readings you got for times 1, 2, and 3 represent the oxygen consumed by endogenous respiration, before you added glucose. For each of the three determinations (tables 3-1, 3-2, 3-3) made by your group, use the graph paper provided at the end of this exercise and your data to graph the values for endogenous respiration as a function of time ($\mu$l $O_2$ consumed on the $y$ axis, time on the $x$ axis). Make three separate graphs.

The readings you got during the rest of each experiment represent the total amount of oxygen used from the beginning of the experiment.

Since total respiration = endogenous respiration + exogenous respiration,

you can calculate exogenous respiration by subtraction if you know how much of the total oxygen consumption was due to endogenous respiration. You didn't measure endogenous respiration at times 4 to 11, but you can estimate these values by extrapolating your curve for times 1-3 (Fig. 3-4, extend line AB to give line BC). Extend the lines on each of the three graphs you made, and use these extrapolations to estimate values for endogenous respiration at times 4 to 11. Enter these values in tables 3-1, 3-2, and 3-3.

After you have estimated the values for endogenous respiration during times 4 to 11, you can estimate oxygen consumption due to exogenous respiration at times 4 to 11 by subtracting estimated endogenous respiration (line BC in Fig. 3-4) from total respiration (line BD in Fig. 3-4). Do this and enter these values on tables 3-1, 3-2, and 3-3. These values tell you how much of the total oxygen consumption was stimulated by the glucose you added.

For each determination done by your group, determine the amount of oxygen consumed during the first two time periods after the addition of glucose by subtracting exogenous respiration at time 4 from exogenous respiration at time 6 and dividing by 6 min. This will give you initial reaction velocities for each of your determinations. Enter these results on the board and in tables 3-4 and 3-5.

Using the data in tables 3-4 and 3-5, plot initial reaction velocities as a function of yeast concentration and as a function of glucose concentration. If there are two or more values for a given concentration, use the average.

Determine final reaction velocities by subtracting exogenous respiration at time 10 from exogenous respiration at time 11 and dividing by 3 min. Again, enter these results on the board and in tables 3-4 and 3-5. Using this information, make plots of final reaction velocity as a function of 1) yeast concentration and 2) glucose concentration.

**Table 3-1.** *Oxygen consumption by yeast, first determination.*

| | | | | μl $O_2$ consumed | | |
|---|---|---|---|---|---|---|

Concentration of yeast:          Glucose concentration:

| Time | (min) | Total endogenous respiration[1] | Total respiration[2] | Total exogenous respiration[3] |
|---|---|---|---|---|
| **Endogenous respiration only** | | | | |
| 0 | 0 | 0 | | - |
| 1 | 5 | | | - |
| 2 | 10 | | | - |
| 3 | 15 | | | - |
| **Endogenous respiration + exogenous respiration** | | | | |
| 4 | 18 | | | |
| 5 | 21 | | | |
| 6 | 24 | | | |
| 7 | 27 | | | |
| 8 | 30 | | | |
| 9 | 33 | | | |
| 10 | 36 | | | |
| 11 | 39 | | | |

[1] Endogenous respiration for times 1-3 measured; endogenous respiration for times 4-11 estimated by extrapolating curve for times 1-3.

[2] Total respiration for times 1-3 = endogenous respiration for times 1-3.

[3] Exogenous respiration for times 4-11 = Total respiration − endogenous respiration.

**Table 3-2.** *Oxygen consumption by yeast, second determination.*

| Concentration of yeast: | | Glucose concentration: | | |
|---|---|---|---|---|
| | | | μl $O_2$ consumed | |
| Time | (min) | Total endogenous respiration[1] | Total respiration[2] | Total exogenous respiration[3] |
| | Endogenous respiration only | | | |
| 0 | 0 | 0 | | - |
| 1 | 5 | | | - |
| 2 | 10 | | | - |
| 3 | 15 | | | - |
| | Endogenous respiration + exogenous respiration | | | |
| 4 | 18 | | | |
| 5 | 21 | | | |
| 6 | 24 | | | |
| 7 | 27 | | | |
| 8 | 30 | | | |
| 9 | 33 | | | |
| 10 | 36 | | | |
| 11 | 39 | | | |

[1] Endogenous respiration for times 1-3 measured; endogenous respiration for times 4-11 estimated by extrapolating curve for times 1-3.

[2] Total respiration for times 1-3 = endogenous respiration for times 1-3.

[3] Exogenous respiration for times 4-11 = Total respiration – endogenous respiration.

**Table 3-3.** *Oxygen consumption by yeast, third determination.*

| Concentration of yeast: | | Glucose concentration: | | |
|---|---|---|---|---|
| | | | $\mu l\ O_2$ consumed | |
| Time | (min) | Total endogenous respiration[1] | Total respiration[2] | Total exogenous respiration[3] |
| Endogenous respiration only | | | | |
| 0 | 0 | 0 | | - |
| 1 | 5 | | | - |
| 2 | 10 | | | - |
| 3 | 15 | | | - |
| Endogenous respiration + exogenous respiration | | | | |
| 4 | 18 | | | |
| 5 | 21 | | | |
| 6 | 24 | | | |
| 7 | 27 | | | |
| 8 | 30 | | | |
| 9 | 33 | | | |
| 10 | 36 | | | |
| 11 | 39 | | | |

[1] Endogenous respiration for times 1-3 measured; endogenous respiration for times 4-11 estimated by extrapolating curve for times 1-3.
[2] Total respiration for times 1-3 = endogenous respiration for times 1-3.
[3] Exogenous respiration for times 4-11 = Total respiration – endogenous respiration.

**Table 3-4.** *Reaction velocities as a function of yeast concentration (pooled class values).*

| Yeast concentration | Initial reaction velocity ($\mu l\ O_2$/min) | Final reaction velocity ($\mu l\ O_2$/min) |
|---|---|---|
| 1 mg/ml | | |
| 2 mg/ml | | |
| 4 mg/ml | | |
| 6 mg/ml | | |
| 8 mg/ml | | |

**Table 3-5.** *Reaction velocities as a function of glucose concentration (pooled class values).*

| Glucose concentration | Initial reaction velocity ($\mu l\ O_2$/min) | Final reaction velocity ($\mu l\ O_2$/min) |
|---|---|---|
| 10 $\mu$M/ml | | |
| 20 $\mu$M/ml | | |
| 30 $\mu$M/ml | | |
| 50 $\mu$M/ml | | |
| 75 $\mu$M/ml | | |
| 100 $\mu$M/ml | | |

## *Questions*

1. What is oxidative phosphorylation?

2. What is a manometer?

3. Why can changes in gas pressure in the reaction vessel be used to measure glucose oxidation?

4. What was the function of the KOH in the reaction vessel?

5. What happens to the number of available active sites as the concentration of substrate in an enzyme-catalyzed reaction increases?

6. What happens to the number of available active sites as the concentration of yeast increases?

7. Describe the relationship between a) initial and b) reaction velocity and glucose concentration and explain the reasons for these relationships.

8. Describe the relationship between a) initial and b) reaction velocity and yeast concentration and explain the reasons for these relationships.

9. What is exogenous respiration?

10. How can $O_2$ consumption due to exogenous respiration be calculated if exogenous respiration is not measured directly?

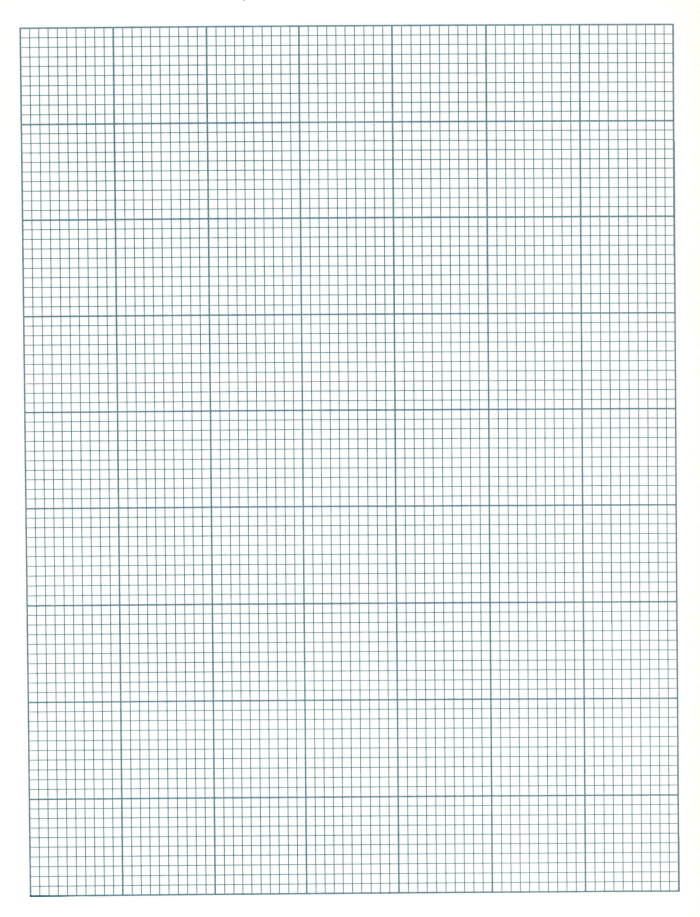

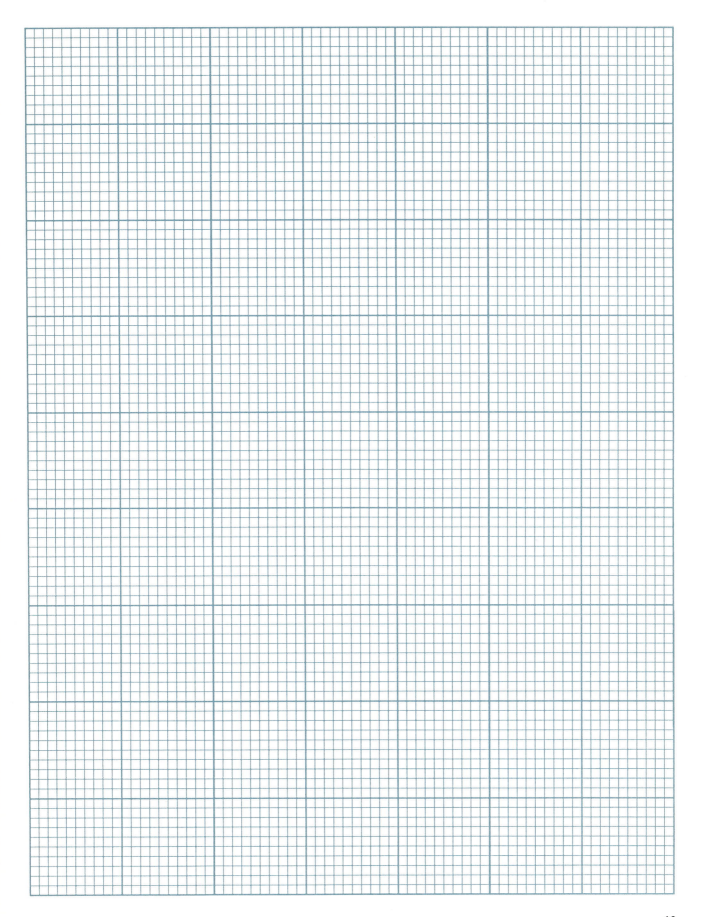

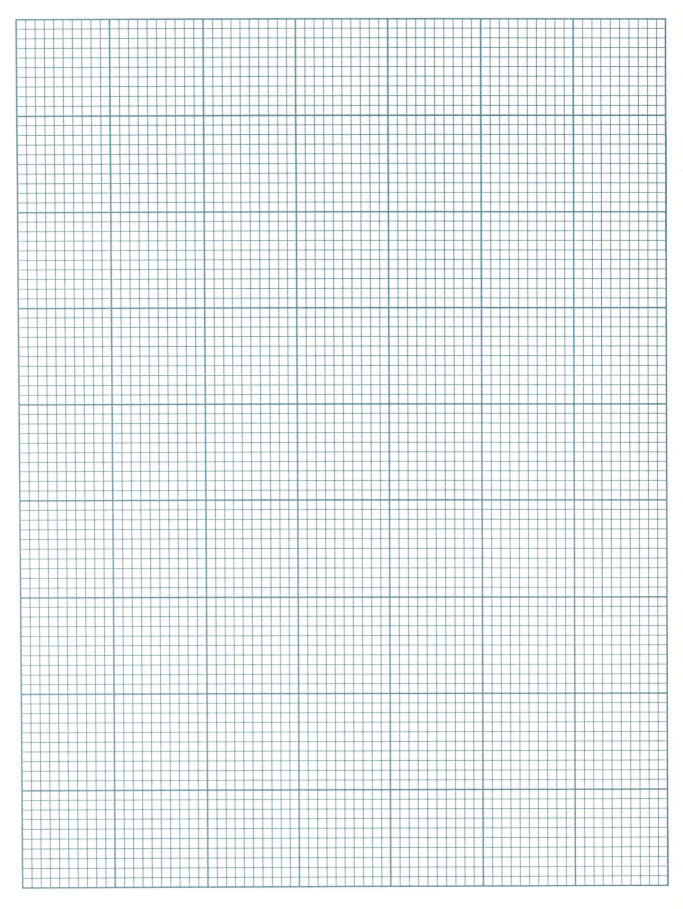

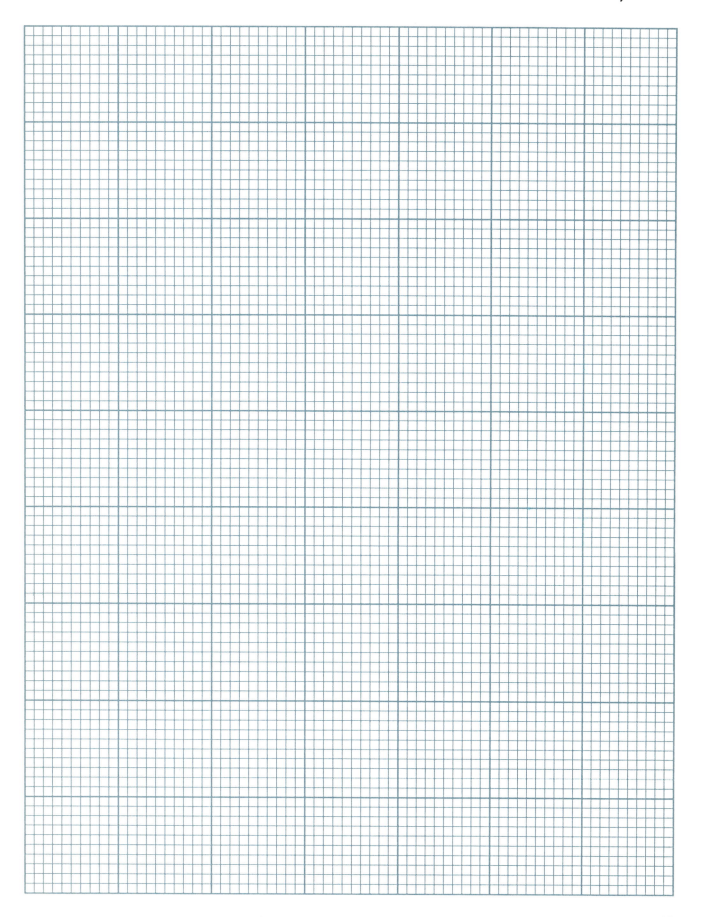

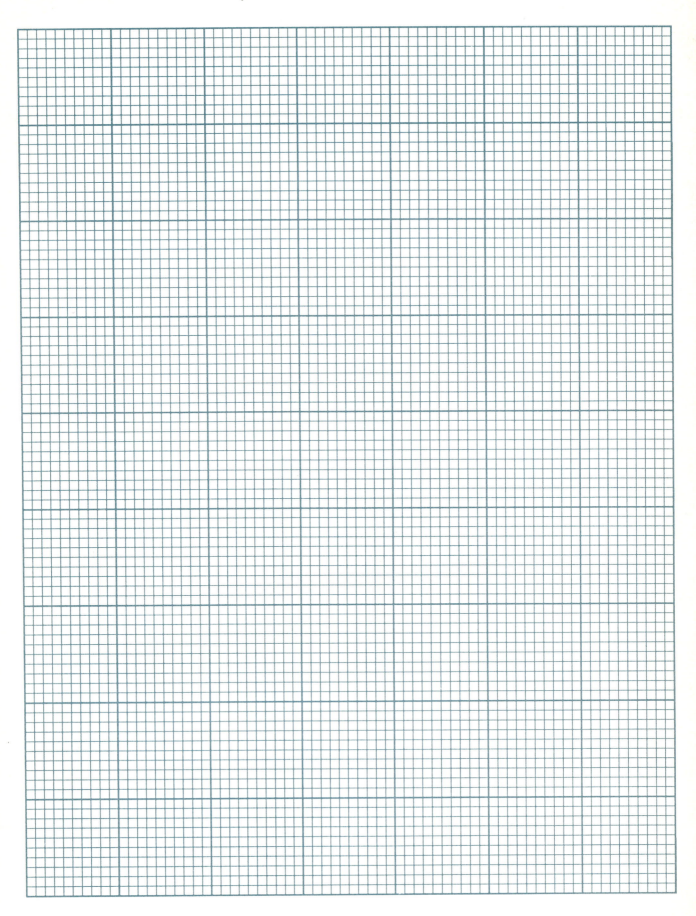

# Diffusion of Solutes

*Reading assignment: text 24-25, 120-125*

## Objectives

### Experimental

1. To determine the effect of molecular weight on diffusion rate.

2. To determine the effect of concentration gradient on diffusion rate.

3. To determine whether various solutions are hypertonic, isotonic, or hypotonic to mammalian red blood cells.

### Conceptual

After completing this exercise and the reading assignment, you should be able to:

1. Define the terms **diffusion, concentration gradient, equilibrium, osmosis, semipermeable, osmotic pressure, molarity, solute, solvent, hyperosmotic, hyposmotic, isosmotic, hypertonic, hypotonic, isotonic, erythrocyte,** and **hemolysis.**

2. Describe two important characteristics of diffusion.

3. List three variables that affect diffusion and explain whether diffusion rate is directly or inversely proportional to each variable.

4. Give five examples of physiological processes that involve diffusion and specify which of these involve osmosis.

5. Describe the effects on cell volume of a solution if its NaCl concentration is specified.

6. Predict the direction(s) in which there will be a net movement of a) water and b) each of several solutes in a system with two compartments separated by a semipermeable membrane, if the substances to which the membrane is permeable are specified and the concentrations of the solutes are given.

## Background

Because of their kinetic energy (see Exercise 1), molecules are in constant motion unless they are at absolute zero (the temperature at which molecular motion ceases). The pathways of molecules vibrating with kinetic energy are random (Fig. 4-1). This random motion is responsible for **diffusion,** the tendency of molecules to move from an area of high concentration to an area of low concentration.

When molecules are distributed uniformly throughout a solution (that is, there are equal concentrations everywhere) they are at **equilibrium** and show no tendency to move preferentially in any direction. However, when concentrations are not the same we say there is a **concentration gradient** between the two regions, and the molecules will diffuse spontaneously from the region of high concentration to the region of lower concentration until they reach equilibrium.

*Diffusion of Solutes*

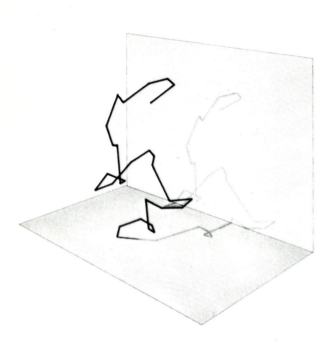

*Figure 4-1. The random path a molecule may take due to its inherent thermal energy.*

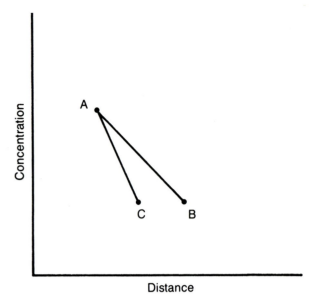

*Figure 4-2. Concentration gradient is a measure of the difference between concentrations of a substance at two points divided by the distance between them.*

Concentration gradient is defined as the difference in the concentrations of a substance at two separate points divided by the distance between the points, or $(C_1 - C_2)/L$, where $C_1$ and $C_2$ are the concentrations at points 1 and 2, and L is the distance between the two points (Fig. 4-2).

If a lump of glucose is placed in a beaker of water (Fig. 4-3A), at first there will be unequal concentrations of glucose in different parts of the vessel (Fig. 4-3B). Initially, some of the glucose will dissolve in the water adjacent to it. This will result in a gradient of glucose in the solution in the beaker (Fig. 4-3B and C). At any given instant, some glucose will move toward the sides of the beaker, some toward the top, and some toward the bottom. Since the glucose concentration is initially high near the bottom of the beaker, at first more glucose molecules will move toward the top of the vessel than toward the bottom. After a while, more molecules will have moved toward the top of the vessel, and the concentration gradient will have diminished. Eventually, the glucose will be uniformly distributed throughout the solution in

the beaker (Fig. 4-3D). When this occurs. there are no longer any differences in concentration, and the system will have reached **equilibrium**.

There are two important characteristics of this process that you should remember:

**\*\* When a difference in concentration exists diffusion will always occur, unless some barrier, like an impermeable membrane, is placed in the way.**

**\*\* This molecular movement is passive and spontaneous. No work need be done to cause diffusion to occur. The energy for the movement is contained in the concentration difference and does not need to be supplied from the outside.**

The rate of diffusion is affected by molecular weight, concentration gradient, and temperature as follows:

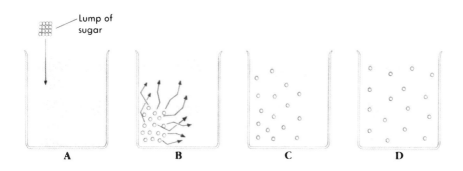

*Figure 4-3. Diffusion.*

** Compounds with high molecular weights diffuse more slowly than compounds with low molecular weights.

** The greater the concentration gradient, the greater the rate of diffusion.

** The higher the temperature of the system, the greater the rate of diffusion.

In the first two parts of this exercise you will investigate the effects of molecular weight (Part 1) and concentration difference (Part 2) on the diffusion of solute molecules. In the third part of today's lab, you will observe evidence of **osmosis**, the diffusion of water through a **semipermeable** membrane. A semipermeable membrane acts as a selective barrier, permitting some substances (in this case, water) to pass through it but not permitting others (in this case, certain solutes) to penetrate. The **osmotic pressure** of a solution is the driving force developed by the difference in water concentration between pure water and a solution (Fig. 4-4).

Although the diffusion of water is denoted by the special name **osmosis**, this process is simply a result of the tendency of molecules to move from a region of higher concentration to a region of lower concentration. Osmosis is a type of diffusion that occurs when something prevents solute molecules from diffusing freely, even though their concentrations differ. For example, a membrane impermeable to solutes will prevent them from moving from a region of high concentration to a region of lower concentration. However, water can move freely through membranes, so it moves into the compartment having the greatest concentration of solute. This movement of water decreases the difference between concentrations on the two sides of the membrane. The concentration of a solution is usually expressed in terms of the concentration of **solute** molecules in the **solvent**. For example, when we refer to a 1 M glucose solution (the number of moles of solute in a liter of solution is termed the **molarity**, abbreviated M) or to a 0.9% NaCl solution, glucose and NaCl are the solutes, and water is the solvent. However, in osmosis, where solute molecules are prevented from moving down their concentration gradients by a barrier, we must look at the concentration of the solvent (water) to determine the direction in which it will move. When solute concentrations differ, so do water concentrations; where solute concentration is low, water is high and vice versa. Water will diffuse from a region where its concentration is high (low solute) to one where its concentration is low (high solute).

To visualize how solute concentration affects water distribution, examine Fig. 4-5. Imagine a beaker containing 100 ml of water divided into two compartments by a semipermeable membrane. If 6 g of a solute to which the membrane is not permeable are added to compartment A and 4 g of the same solute are added to compartment B, the concentration of solute will now be higher in A (which has 6 g in 50 ml) than in B (which has 4 g in 50 ml) (Fig. 4-5A). Consequently, the concentration of water will be higher in side B than in side A,

*Figure 4-4. Osmotic pressure can be measured by placing a solution into a bag made of a membrane that is permeable to water but not to solute (A), connecting a vertical tube to the bag, and measuring the height of the solution in the tube (B).*

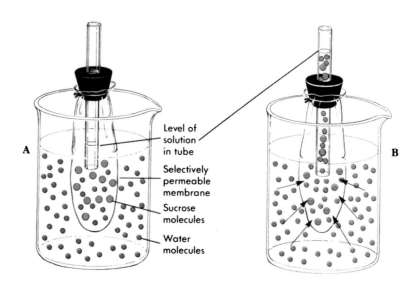

and water will move by osmosis from B to A until the concentrations of the solutions in the two compartments are equal (Fig. 4-5B). (Assume that the effect of atmospheric pressure on the height of the column is insignificant.)

Diffusion and osmosis are important processes in the body. Some (but not all) molecules are transfered from the environment into the blood (for instance, in the intestine) or from blood into the environment (as in the lungs) by diffusion. In addition, diffusion causes the movement of some molecules from the blood into a cell or from a cell into the blood. This transfer of molecules is very effective over extremely short distances (a fraction of a mm or less). However, over longer distances (e.g., cm) diffusion is too slow to deliver much material.

Osmosis is responsible for the absorption of much water from the intestine and in the kidney (which would otherwise produce urine at an embarrassing rate!). On the other hand, if the concentration of blood proteins decreases (for instance, as a result of liver or kidney disease or severe malnutrition) the osmotic pressure of the blood decreases, and the result is the movement of water out of the blood and accumulation of fluid in the tissues (**edema**).

If the concentration of solute molecules surrounding a cell is greater than the fluid inside the cell, the external fluid is said to be **hyperosmotic** to the cell. If the solute concentration in the external solution is lower than the concentration of solutes inside the cell, the cell is in a **hyposmotic** solution. If solute concentration is the same inside and outside of the cell, the cell is in an **isosmotic** solution. The concentration of sodium chloride in red blood cells (**erythrocytes**) is approximately 0.9% NaCl; consequently, a 0.9% NaCl solution would be isosmotic to red blood cells. (This information will be useful to you when you formulate predictions for the experiments you will do in this exercise.) Note that when comparing osmotic pressures, it is important to be clear about the point of reference. For instance, if a cell is in an hyperosmotic solution, that is, the external solution is hyperosmotic to the cell, then the cell must be hyposmotic to the solution.

Three additional terms describe the effects of solutions on cell volume. If a cell shrinks when it is placed in a solution, the solution is **hypertonic** to the cell, and if there is no change in cell volume, the surrounding solution is **isotonic**. If a cell swells (and perhaps bursts) in a solution, then the solution is **hypotonic** to the cell. In Exercise 5 you will see that the effect ions in an extracellular solution have on a cell depends on whether or not

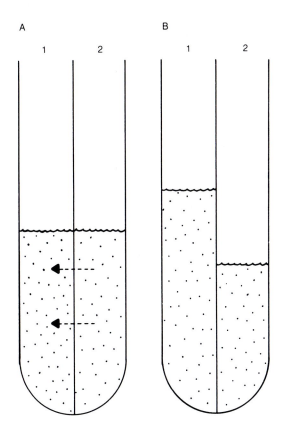

A              B

1      2      1      2

*Figure 4-5. Effect of solute concentration on water distribution when a membrane is permeable to water but not to solute.*

the cell membrane is permeable to the solutes in question. For instance, an isosmotic solution is not necessarily isotonic. This is why it is necessary to have two separate sets of terms, one to compare osmotic pressures inside and outside the cell and one to describe a solution's effect on cells.

When red blood cells are placed in a hypotonic solution, the bursting of cells that results is termed **hemolysis**. A suspension of intact red blood cells will scatter light passing through it; however, erythrocytes placed in a hypotonic solution will swell and burst and will, therefore, fail to scatter light. No image is formed on your retina when you view newsprint through a suspension of intact red blood cells, but a clear image appears when newsprint is examined through a suspension in which hemolysis has taken place. In Part 3 of this exercise, you will take advantage of this fact

to determine whether or not test solutions are hypotonic to red blood cells.

Before beginning this exercise, be sure you understand the hypotheses you will be testing and the predictions that stem from them.

**Hypothesis 1:** The rate of diffusion of solute molecules decreases as molecular weight of the solute increases.

**Hypothesis 2:** The rate of diffusion of solute molecules increases as the concentration gradient increases.

**Hypothesis 3:** Osmosis causes the movement of water into cells when they are placed in hyposmotic conditions and out of cells placed in hyperosmotic solutions; however, there is no net movement of water in either direction when cells are placed in isosmotic conditions.

To be sure you understand how osmosis and diffusion work, complete these predictions.

**Prediction 1:** The order of the diffusion rates for the substances methylene blue, alcian blue, ruthenium blue, and methyl red (from the fastest to the slowest) will be

**Prediction 2:** The order of the diffusion rates for solutions of 0.3 mM methylene blue, 1 mM methylene blue, and 3 mM methylene blue (from the fastest to the slowest) will be

_____.

**Prediction 3:** When red blood cells are placed in 0.9% NaCl, they will

_____.

**Prediction 4:** When red blood cells are placed in distilled water, they will

_____.

**Prediction 5:** When red blood cells are placed in 10% NaCl, they will

_____.

## Procedure

Record your results in the laboratory report section at the end of this exercise.

 *Part 1. The effect of molecular weight on diffusion rate*

1. In this exercise you will use solutions of four different compounds at the same concentration (1 mM). All of these compounds are brightly colored. This makes it possible to follow the progress of the molecules as they diffuse in a translucent medium. The four compounds have different molecular weights; their names and molecular weights are:

> methyl red (269 g/M),
> methylene blue (374 g/M),
> ruthenium red (786 g/M),
> alcian blue (1,300 g/M).

2. Obtain four test tubes, each containing 15 ml of agarose, a gel. This is essentially a column of water. The only purpose of the agarose is to make the column rigid, so it won't mix with the other solutions (see below). Add 5 ml of one of the dye solutions to the top of each agarose column. Note that the interface between the colored solution and the agarose is sharp; no mixing occurs.

3. Record the time in Table 4-1. Observe that the dye molecules begin to move into the agarose as time passes.

4. At the end of the lab period measure the distance in mm between the top of the agarose and the farthest point at which you can see the color. Record both the distance and the time of your observation (Table 4-1).

5. Repeat the measurement 6 to 12 hr after the end of your laboratory period and then once each day for two more days. Record your observations in Table 4-1.

*Part 2. The effect of concentration gradient on diffusion rate*

You will compare the diffusion rates of three solutions of methylene blue at different concentrations (0.3 mM, 1 mM, and 3 mM).

1. Obtain four test tubes containing agarose.

2. Add the different methylene blue solutions to the top of the agarose column.

3. Make the same measurements as you made in the previous section, and record your observations in Table 4-2.

 *Part 3. Water diffusion (osmosis)*

1. Obtain a suspension of mammalian blood cells in Ringer's solution. Several glass slides and cover slips, test tubes, and a microscope will be provided. You will also need the three solutions listed below:

> distilled water, which is less concentrated than (hyposmotic to) Ringer's solution;
>
> 0.9% NaCl, which has the same total solute concentration as (is isosmotic with) Ringer's;
>
> 10% NaCl, which is much more concentrated than (hyperosmotic to) Ringer's solution.

2. Place a small drop of cell suspension on a slide and drop a cover slip on it. Hold it over some print. Record your observations in Fig. 4A-I.

3. Place a few drops of the blood suspension on two other slides. To one of them add three drops of 0.9% NaCl and mix. To the other add three drops of distilled $H_2O$ and mix. Hold them above a printed page and try to read the print through the solutions. Sketch your observations in figs. 4-B-I and 4-C-I.

4. In the last section you were asked to deduce, from indirect evidence, something that happened to one of the blood cell suspensions. If your deduction was correct it should be possible to repeat the experiment, slightly modified, so that you can watch the suspension under a microscope and observe the event directly. To do this, place another drop of cell suspension on a slide and add a cover slip. Observe under the low power lens of the microscope. Record your observations (Fig. 4-A-II and III).

5. While you are watching, your partner will place several drops of 0.9% NaCl around the edges of the cover slip. The cells will move around as water streams in, but disregard this. Make sketches of your observations in Fig. 4-B-II and III.

6. Make another slide with a drop of cell suspension and a cover slip. This time apply distilled water around the edges. Record your observations (Fig. 4-C-II and III).

7. Place several ml of 10% NaCl solution in a test tube, add 3 drops of cell suspension, and mix thoroughly. After several minutes add a drop of this mixture to a slide, add a cover slip, and observe the cell under low and high power lenses of the microscope. Record in Fig. 4-D. Compare the appearance of these cells to the appearance of normal erythrocytes. (If necessary, observe some cells suspended in 0.9% NaCl solution for comparison).

Name: _____

Date: _____

Lab Section: _____

## Analyzing Your Data

If other variables, such as concentration and temperature, are equal, your results in Table 4-1 would be expected to show how molecular size affects diffusion. Using the data for times 1, 2, and 3 from Table 4-1, prepare a graph showing the effect of molecular weight on diffusion rate. Molecular weight is your independent variable, so graph it on the abscissa (*x*-axis), and graph diffusion rate, your dependent variable, on the ordinate (*y*-axis). Remember, rate refers to a process that occurs per unit time, so you need to convert the values in Table 4-1, which are distances in mm, to mm moved per minute, or mm/min. Be sure to label and give units for each axis.

Your graph should include three lines, one each for times 1, 2, and 3. Distinguish the three lines, either by using different colors or with different kinds of lines (for example, solid, dotted, and dashed lines). Include a legend which explains what the different colors or line types represent.

Next, use the data you recorded in Table 4-2 to determine the effect of concentration on diffusion rate. Make another graph, but this time use the data from times 1, 2, and 3 in Table 4-2, to depict the relationship between concentration and diffusion rate.

To analyze your data on osmosis, consult your notes and the diagrams you made (figs. 4-A to 4-D). Be sure that you understand in which direction there was a net movement of water and why before you begin answering the questions below.

*Diffusion of Solutes*

*Table 4-1. Effect of molecular weight on rate of diffusion.*

**Observation times:**

| Time 1: | Day 1: | Start of experiment: |
|---|---|---|
| Time 2: | Day 1: | End of lab: |
| Time 3: | Day 1: | 6 to 12 hr after lab: |
| Time 4: | Day 2: | |
| Time 5: | Day 3: | |

Distance moved by solute molecules

| Time | Time elapsed (min) | Methyl red (269 g/M) | Methylene blue (374 g/M) | Ruthenium red (786 g/M) | Alcian blue (1,300 g/M) |
|---|---|---|---|---|---|
| 1. | | | | | |
| 2. | | | | | |
| 3. | | | | | |
| 4. | | | | | |
| 5. | | | | | |

*Table 4-2. Effect of concentration on rate of diffusion.*

**Observation times:**

| Time 1: | Day 1: | Start of experiment: |
| Time 2: | Day 1: | End of lab: |
| Time 3: | Day 1: | 6 to 12 hr after lab: |
| Time 4: | Day 2: | |
| Time 5: | Day 3: | |

| | | Concentration of methylene blue | | |
| --- | --- | --- | --- | --- |
| Time | Time elapsed (min) | 0.3 mM | 1 mM | 3 mM |
| 1. | | | | |
| 2. | | | | |
| 3. | | | | |
| 4. | | | | |
| 5. | | | | |

I. Appearance of newsprint      II. Appearance of cells      III. Abundance of cells

*Fig. 4-A. Control: Untreated blood cell suspension.*

I. Appearance of newsprint      II. Appearance of cells      III. Abundance of cells

*Fig. 4-B. Blood cells in 0.9% NaCl.*

I. Appearance of newsprint     II. Appearance of cells     III. Abundance of cells

*Fig. 4-C. Blood cells in distilled water.*

I. Appearance of newsprint     II. Appearance of cells     III. Abundance of cells

*Fig. 4-D. Blood cells in 10% NaCl.*

## *Questions*

1. What causes diffusion?

2. Choose the correct answer: At a given temperature, the diffusion rate of large molecules is (greater than or less than) the diffusion rate of small molecules. _____ Explain.

4. Under which conditions: a) undiluted suspension of red blood cells, b) red blood cells in 0.9% NaCl, or c) red blood cells in distilled water were you able to read newsprint through a suspension of red blood cells? _____ Explain.

5. Were you able to distinguish individual cells when you viewed the red blood cell suspension under the microscope?

_____

6. Describe what happened to the appearance and the abundance of the red blood cells under each of the following conditions:

when you added 0.9% NaCl     _____

when you added distilled water     _____

when you added 10% NaCl     _____
Explain.

7. Indicate whether each of the solutions below is isotonic, hypertonic, or hypotonic to mammalian red blood cells:

0.009% NaCl     _____

5% NaCl     _____

0.9% NaCl     _____

8. Indicate whether the intracellular fluid of mammalian red blood cells is isosmotic, hyperosmotic, or hyposmotic to each of the solutions below.

10% NaCl     _____

distilled water     _____

0.01% NaCl     _____

9. How might diffusion rates affect the maximum size cells can attain? (Hint: why is the maximum distance between a cell and the nearest capillary typically no greater than 0.01 mm?)

10. Choose the correct answer: A frog living in a freshwater pond is in an environment that is (hypertonic, isotonic, hypotonic) to the frog's tissues. _____

11. The diagrams in Figs. 4-E to F below depict two fluid-filled compartments separated by a membrane of specified permeability. For each substance shown in each of Figs. 4-E to F below, indicate whether there will be a net movement of the substance and, if so, the direction of the movement. (Assume the membrane is permeable to water in all cases.)

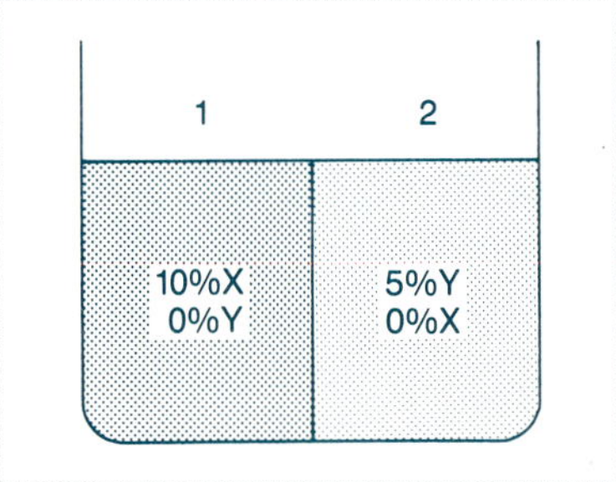

| 1 | 2 |
|---|---|
| 10%X 0%Y | 5%Y 0%X |

*Figure 4-E.*

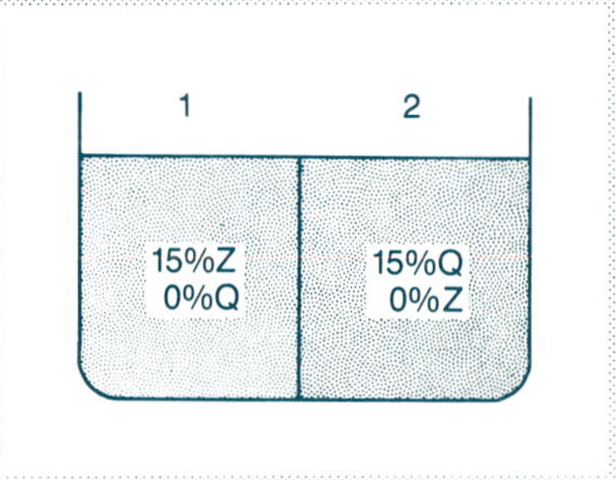

| 1 | 2 |
|---|---|
| 15%Z 0%Q | 15%Q 0%Z |

*Figure 4-F.*

a) Membrane equally permeable to X and to Y. (Assume that the molecular weight of X is equal to the molecular weight of Y.)

Water _____

X _____

Y _____

b) Membrane permeable to Q but not to Z.

Water _____

Q _____

Z _____

12. a) Assume that you are a health care practitioner who is treating an accident victim. The patient has lost a good deal of blood, and you decide to administer an intravenous infusion to restore blood volume and blood pressure. What concentration of NaCl would be best for your patient? _____

b) What will be the consequences for the volume of intracellular fluid if the solution you administer is too dilute?

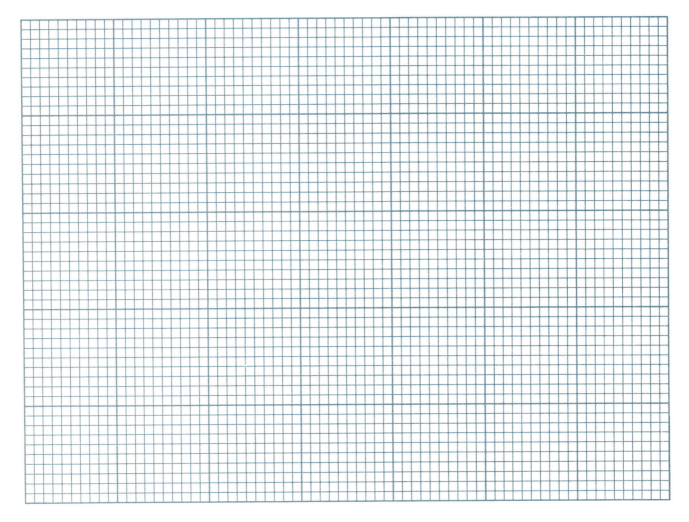

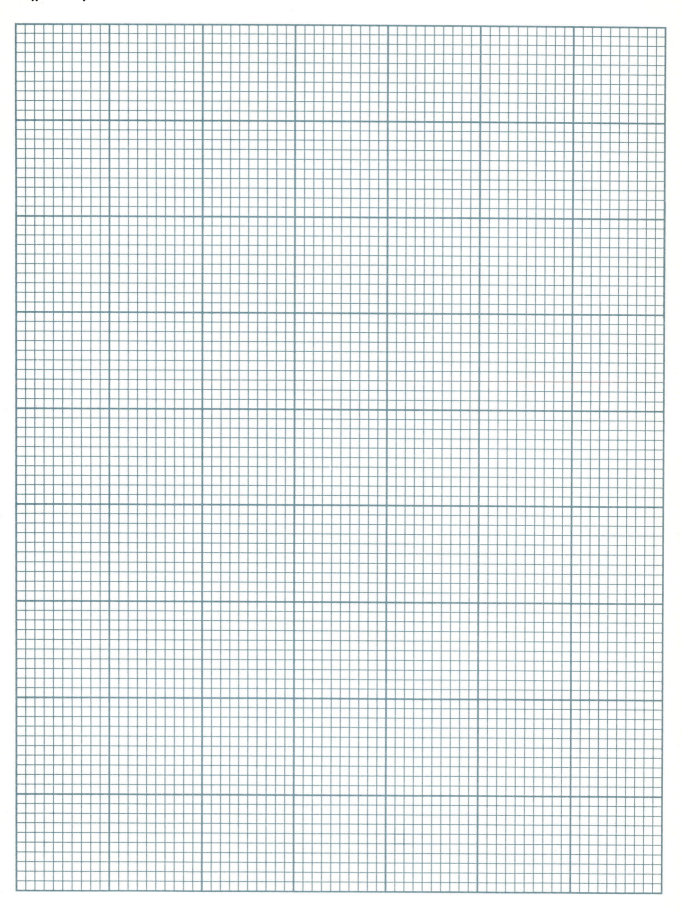

# 5 Permeability of Cell Membranes

*Reading assignment: text 117-120, 123-128*

## Objectives

### Experimental

1. To determine the relationship between hemolysis time and concentration of sodium chloride solution.

2. To determine the relationship between hemolysis time and molecular weight for water soluble compounds.

3. To determine the relationship between hemolysis time and lipid solubility for nonpolar compounds.

### Conceptual

After completing this exercise and the reading assignment, you should be able to:

1. List two factors that regulate the passive movement of substances across cell membranes.

2. Discuss the difference between the driving force for charged and for uncharged molecules.

3. Describe the structure of cell membranes according to the fluid mosaic model.

4. Define the terms **phospholipid, polar, nonpolar, hydrophilic, hydrophobic,** and **lipid bilayer.**

5. Distinguish between membrane channels and membrane carriers.

6. Compare the mechanisms by which water soluble and lipid soluble compounds cross cell membranes.

7. Explain how hemolysis can be detected in the laboratory without the use of special equipment.

8. Explain why hemolysis time can be used as an index of membrane permeability.

9. Explain how hemolysis can occur when a suspension of red blood cells is placed in an isosmotic solution.

## Background

The passive movement of materials across cell membranes is the result of a) the **permeability** of the membrane to the diffusing substance and b) the **driving force**. For uncharged compounds, the driving force is a **concentration gradient** (see Exercise 4) between the inside and the outside of the cell. In the case of ionic diffusion, the driving force results from the concentration gradient and the **electrical potential difference** between the cell's interior and its exterior.

In general, biological membranes are far less permeable than an equivalent thickness of solvent, and they are completely impermeable to many compounds. Membrane permeability, active transport, and driving force determine the rate of movement of nutrients into and waste compounds out of the cell. The penetration rate of a cell

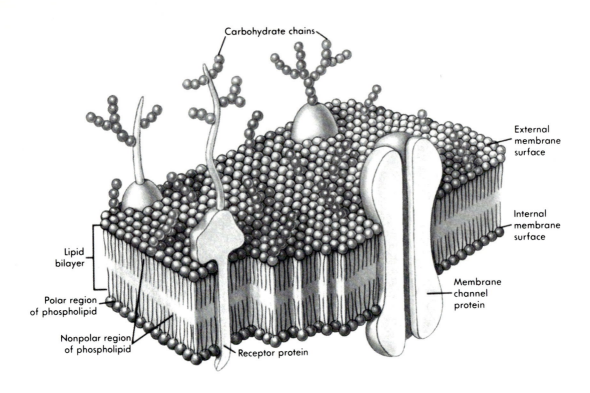

**Figure 5-1.** *Fluid mosaic model of a cell membrane composed of a lipid bilayer with proteins "floating" in the membrane. The nonpolar hydrophobic portion of each phospholipid molecule is directed toward the center of the membrane, and the polar hydrophilic portion is directed toward the water environment either outside or inside the cell. Intrinsic proteins partially or fully penetrate the membrane and may serve as transport or receptor molecules.*

membrane by a molecular species via diffusion is important in determining the chemical composition of the cell. A substance to which the membrane is impermeable will not be found in the cytoplasm unless it is produced there or actively transported into the cytoplasm. In this exercise, we will investigate the effects of lipid solubility and molecular weight on the permeability of red blood cell membranes.

The structure of a cell membrane is shown in Fig. 5-1. This representation is referred to as the **fluid mosaic model**, because the membrane is a mosaic of phospholipids and proteins, and, in many cases, these constituents are free to move within the membrane.

The cell membrane is a thin membrane of **phospholipids** studded with proteins. Phospholipids consist of two fatty acid chains and one **polar** (charged) phosphate compound attached to a glycerol backbone (Fig. 5-2). When they are surrounded by water molecules, the **hydrophilic** (meaning "water-loving") ends of the phospholipid molecules face outward toward the water, while the **hydrophobic** (water-avoiding) hydrocarbon chains point away from the water. In the membranes of cells and organelles, this results in two layers of phospholipids, called a **lipid bilayer**. The hydrocarbon interior of the membrane constitutes a barrier to penetration of polar solutes, such as amino acids, sugars, and ions.

The mechanism by which a solute in the extracellular fluid enters a cell depends on the polarity of the solute. Nonpolar solutes pass directly through the membrane's lipid matrix. Therefore, within a group of chemically related, nonpolar compounds, those with high lipid solubility will enter the cell more readily than those with low lipid solubility.

*Figure 5-2. A Phospholipids, composite molecules similar to a triglyceride, are formed by replacing one of the fatty acids of a triglyceride with a polar phosphate compound (here choline addition forms lecithin or phosphatidyl choline).*
*B Phospholipids are usually oriented so that the polar portion extends from one end of the molecule and the two fatty acid chains from the other, so they are often represented diagramatically as a polar ball with a nonpolar tail.*

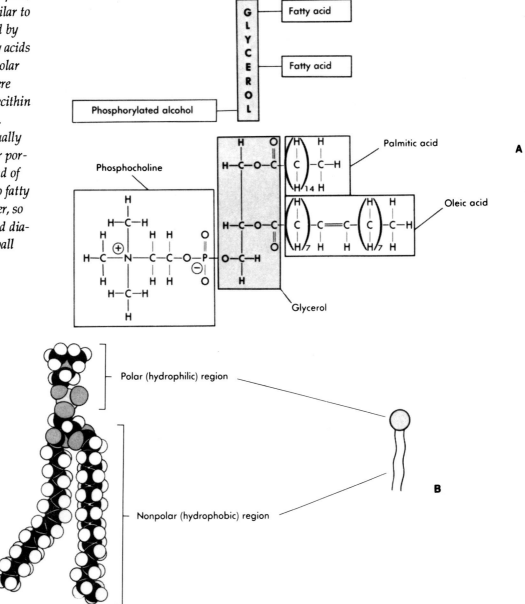

Most of the substances a cell comes in contact with have charged groups, and they are water soluble. Therefore, they cannot pass through the lipid portions of membranes. These polar solutes must take specific avenues through lipid membranes. Molecules of low lipid solubility can pass through pores in the lipid membrane if they are small. On the other hand, large molecules of low lipid solubility will be excluded.

Some solutes that are not lipid soluble can penetrate the cell membrane with the aid of proteins in the membrane. The simplest way for a polar solute to cross a cell membrane is by way of pores, or **channels** (Fig. 5-3A). A channel is a tube through a membrane formed by one or more intrinsic membrane proteins. On the outside of the tube, facing the phospholipid, are many hydrophobic groups. The inside of the tube is lined with hydrophilic groups, allowing the channel to be filled with water. Thus, polar solutes can diffuse through the membrane in aqueous solution. A more complicated mechanism by which polar solutes enter cells involves **membrane** carriers, which undergo a cycle of binding and conformational change as solutes move across the membrane (Fig. 5-3B).

To summarize, although the permeability of membranes is highly variable among organisms, two generalizations seem to be valid for all cells:

**\*\* For chemically related water soluble compounds, the rate of diffusion across cell membranes decreases as molecular size increases (that is, small molecules enter the cell faster than large molecules).**

**\*\* For chemically related, nonpolar compounds, the rate of diffusion across cell membranes increases with increasing lipid solubility and is (more or less) independent of molecular size.**

Because cell membranes are selectively permeable, we need to distinguish between the osmotic pressure of a solution and its effect on cells (see Exercise 4). Imagine that a cell is placed in a solution having a total solute concentration equivalent to the concentration of solutes within the cell (Fig. 5-4). The cell is in an isosmotic solution.

*Figure 5-3. Comparison of the operations of membrane channels and membrane carriers.* **A** *A channel is a tube through the membrane formed by one or more intrinsic membrane proteins.* **B** *In contrast to channels, carriers must undergo a cycle of binding and conformational change.*

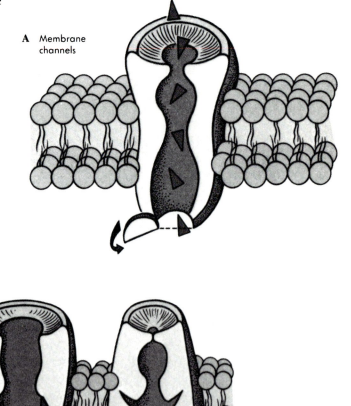

A   Membrane channels

B   Membrane carriers

1           2           3

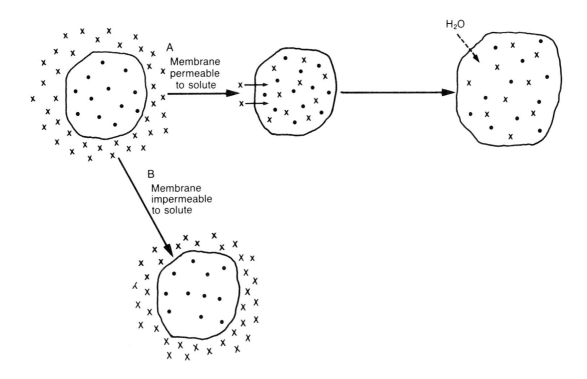

*Figure 5-4. The fate of a cell placed in an isosmotic solution depends on whether or not the cell membrane is permeable to the solutes in the extracellular solution. If the membrane is not permeable to the solute there will not be a net movement of water either into or out of the cell, and the cell will neither shrink nor swell (A). Water will follow solute into the cell, and the cell will swell (B).*

However, if the solute molecules in the external solution are not found, or are found in lower concentrations, inside the cell, the effect of solute molecules in the solution surrounding a cell will depend on whether or not the membrane is permeable to them. If the membrane is not permeable to solute, there will be no net movement of water into or out of the cell (Fig. 5-4A). Such a cell is in a solution that is isotonic. However, if the solute can cross the membrane, it will enter the cell, because of the higher concentration of solute outside the cell (Fig. 5-4B). When solute enters, the osmotic pressure inside the cell increases, and water enters the cell. This influx of water causes the cell to swell, and it may burst. In this case the cell is in a hypotonic solution (see Exercise 4), even though the solution is isosmotic.

One of the best known preparations for studying membrane permeability is the mammalian erythrocyte. In the previous exercise you saw that hemolysis occurs when red blood cells are placed in distilled water. For the reasons given in the preceding paragraph, hemolysis can also take place when a solute in the extracellular solution moves down its concentration gradient and enters a cell, increasing the internal osmotic pressure.

When hemolysis occurs, the internal solutes, including hemoglobin, escape. This is the basis of our method for following the diffusion of solutes from the external bathing medium into the cell. Recall that a suspension of blood cells (in plasma or in physiological saline) scatters light (see Exercise 4). A sharp image of a small object viewed through the suspension cannot be formed. Consequently, a thread or a piece of newsprint viewed through a suspension of blood cells will appear faint and fuzzy; it may not be seen at all if the suspension is dense. However, when the cells hemolyze, the solution becomes translucent, and a sharp image can be seen (Fig. 5-5).

**The faster a solute diffuses into the cell, the sooner it will hemolyze.** Thus, the time taken for the thread to appear gives us a measure of membrane permeability to the solute in question, providing that the driving force remains the same. We will take advantage of this process to compare the permeability of the erythrocyte membrane to different solutes and to water.

In the first part of this exercise you will determine what concentration of NaCl is isosmotic to erythrocytes. In parts 2 and 3, you will use compounds that have been prepared in isosmotic solution (that is, the solutions have the same osmotic pressure as the cell interior). You will work with water soluble compounds in Part 2 to investigate the relationship between molecular weight and hemolysis time for polar compounds, and in Part 3 you will investigate the relationship between lipid solubility and hemolysis time.

Before beginning the experiments that follow, be sure you understand why hemolysis time can be used as an indicator of membrane permeability and how solutes in an isosmotic solution can cause hemolysis to occur.

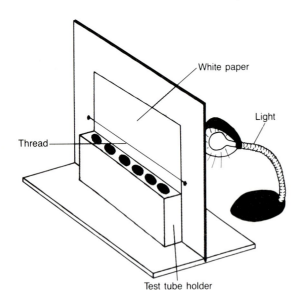

*Figure 5-5. Apparatus for measuring time to hemolysis.*

## Procedure

Record your results in the laboratory report section at the end of this exercise.

Throughout this exercise, repeat each of your measurements three times and use the average of the three measurements as your hemolysis time.

 *Part 1. Permeability to water*

1. Mount the hemolysis apparatus in front of a lamp. (Your instructor will explain the use of the apparatus.) Arrange a series of six small vials in the rack on the front.

2. NaCl solutions with the following concentrations will be available:

> 0.10 M,
> 0.09 M,
> 0.08 M,
> 0.07 M,
> 0.06 M,
> 0.05 M.

To each tube in the hemolysis apparatus add 2 to 3 ml of one of the NaCl solutions as shown in Figure 5-6.

3. Noting the <u>exact</u> time of addition (record this in Table 5-1), add two drops of a well-mixed blood suspension to the first tube, mix rapidly, and replace on the rack. If the thread becomes visible, record the time of its appearance (Table 5-1), and consider the time elapsed to be the hemolysis time.

4. Repeat with each of the other tubes. Do not wait for more than a minute or so; if hemolysis has not occurred within that time, it probably won't occur at all.

5. Repeat steps 1 to 4 two more times.

Test tubes in hemolysis apparatus

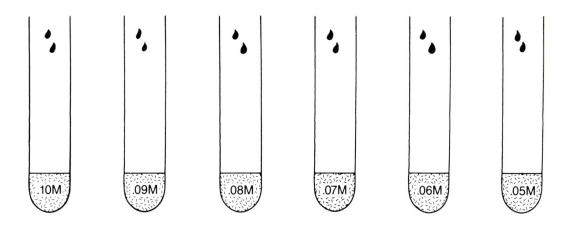

2-3 ml NaCl solution

blood suspension

*Figure 5-6. Experimental set-up for measuring hemolysis time of red blood cells in NaCl solutions.*

 ### Part 2. Permeability to water soluble compounds

1. Mount another series of vials in the rack. Into each vial place one of the following solutions (Fig. 5-7):

> 0.3 M methanol,
> 0.3 M ethylene glycol,
> 0.3 M glycerol,
> 0.3 M erythritol.

2. Using the same technique as above, determine hemolysis time in each of the first four vials. Record your results in Table 5-2.

 ### Part 3. Permeability to nonpolar compounds

1. Using the same technique as above, measure the permeability of the red blood cell membrane to 0.3 M solutions of the following acetate esters of glycerol (Fig. 5-8):

> glycerol,
> monacetin,
> diacetin,
> triacetin.

Note that lipid solubility increases from glycerol through triacetin. Molecular size varies in the same way: glycerol is the smallest molecule; triacetin is the largest.

2. Record your results in Table 5-3.

Test tubes in hemolysis apparatus

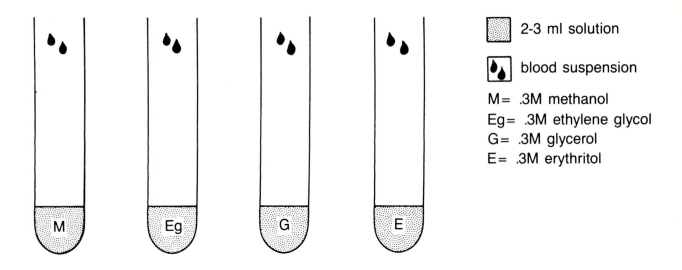

**Figure 5-7.** *Experimental set-up for measuring hemolysis time of red blood cells in solutions of water soluble compounds.*

Test tubes in hemolysis apparatus

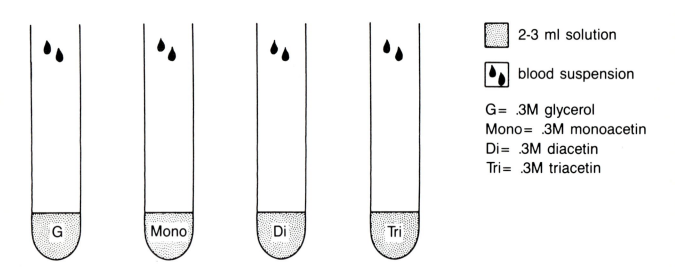

**Figure 5-8.** *Experimental set-up for measuring hemolysis time of red blood cells in solutions of nonpolar compounds.*

# Laboratory Report

## *Exercise 5: Permeability of Cell Membranes*

*Name:* ______________________

*Date:* ______________________

*Lab Section:* ______________________

## *Analyzing Your Data*

The units of hemolysis time are sec. The reciprocal of hemolysis time will be per unit time (1/sec or $sec^{-1}$). This is a rate. It is your measure of permeability.

Use your data from Table 5-1 to plot average hemolysis time as a function of the concentration of NaCl in the extracellular solution.

For the data in Table 5-2, plot the relationship between average hemolysis time and molecular weight.

***Table 5-1.*** *Relationship between hemolysis time and concentration of sodium chloride solution.*

| Concentration of NaCl solution | Time at addition of blood suspension a[1] b c | Time at appearance of thread a b c | Time to hemolysis (sec) a b c | Average hemolysis time (sec) |
|---|---|---|---|---|
| 0.10 M | | | | |
| 0.09 M | | | | |
| 0.08 M | | | | |
| 0.07 M | | | | |
| 0.06 M | | | | |
| 0.05 M | | | | |

[1] a, b, and c denote first, second, and third trials.

To analyze the data in Table 5-3, it will be helpful if you prepare two graphs:

1) Plot average hemolysis time as a function of molecular weight.

2) Prepare a bar graph (see Appendix 1) with relative lipid solubility on the *x*-axis and hemolysis time on the *y*-axis.

The **oil/water partition coefficient** is a measure of lipid solubility. Relative values for this term are given in Table 5-3. To determine this value, the compound in question is shaken in a mixture of equal amounts of water and olive oil, and the solubilities in lipid and water are measured:

$$\frac{\text{partition}}{\text{coefficient}} = \frac{\text{solute concentration in lipid.}}{\text{solute concentration in water}}$$

*Table 5-2. Relationship between hemolysis time and molecular weight for water soluble compounds.*

|  | Mol. Wt. (g/M) | Time at addition of blood suspension a[1] b c | Time at appearance of thread a b c | Time to hemolysis (sec) a b c | Ave. hem. time (sec) |
|---|---|---|---|---|---|
| 0.3 M Methanol | 32 |  |  |  |  |
| 0.3 M Ethylene glycol | 62 |  |  |  |  |
| 0.3 M Glycerol | 92 |  |  |  |  |
| 0.3 M Erythritol | 122 |  |  |  |  |

[1] a, b, and c denote first, second, and third trials.

*Table 5-3.* Relationship between hemolysis time and lipid solubility for nonpolar compounds.

| | Lipid Sol.[1] | Mol. Wt. (g/M) | Time at addition of blood susp. a[2] b c | Time at appear. of thread a b c | Time to hemol. (sec) a b c | Ave. hem. time (sec) |
|---|---|---|---|---|---|---|
| 0.3 M Glycerol | low | 92 | | | | |
| 0.3 M Monoacetin | medium | 134 | | | | |
| 0.3 M Diacetin | high | 176 | | | | |
| 0.3 M Triacetin | very high | 218 | | | | |

[1] Based on oil/water partition coefficient.

[2] a, b, and c denote first, second, and third trials.

## *Questions*

1. Why can we use hemolysis time as an indication of membrane permeability?

2. In Part 1, at which concentration(s) of NaCl did hemolysis occur?
_____

3. What caused solutes to enter the red blood cells in Part 1?

4. Can hemolysis occur when erythrocytes are placed in an isosmotic solution? _____ Explain.

5. Indicate whether or not each of the solutions listed below was hypotonic to mammalian erythrocytes:

0.3 M methanol          _____

0.3 M ethylene glycol   _____

0.3 M glycerol          _____

0.3 M erythritol        _____

0.3 M monoacetin        _____

0.3 M diacetin          _____

0.3 M triacetin         _____

6. In Part 2, how did the water soluble solutes enter the red blood cells?

7. How did the nonpolar solutes enter the red blood cells in Part 3?

8. Many organic solvents, such as benzene (a component of paint thinners), are highly toxic. Apply what you have learned in this lab to explain the basis of this toxicity.

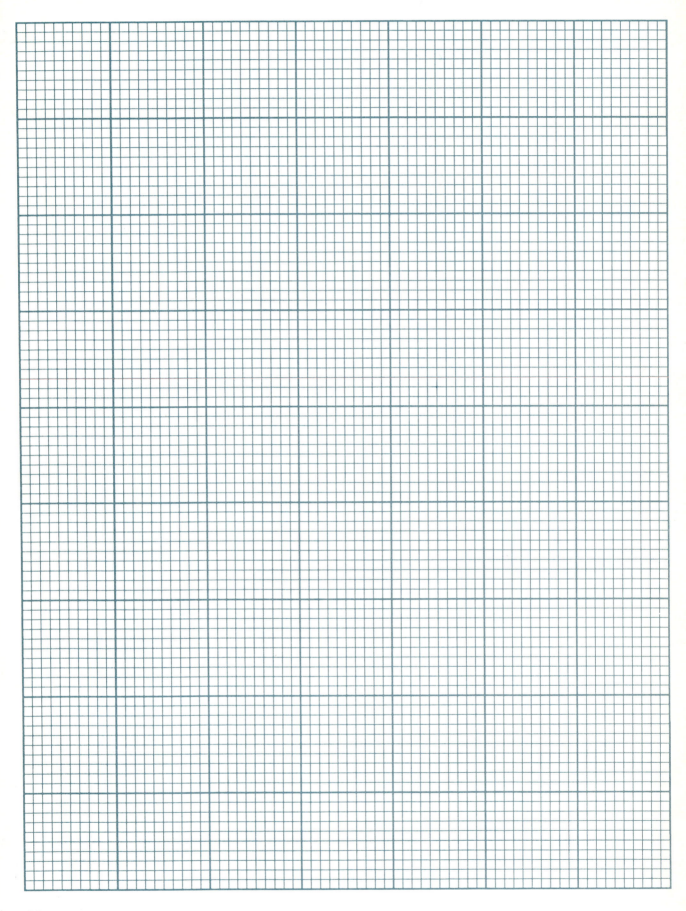

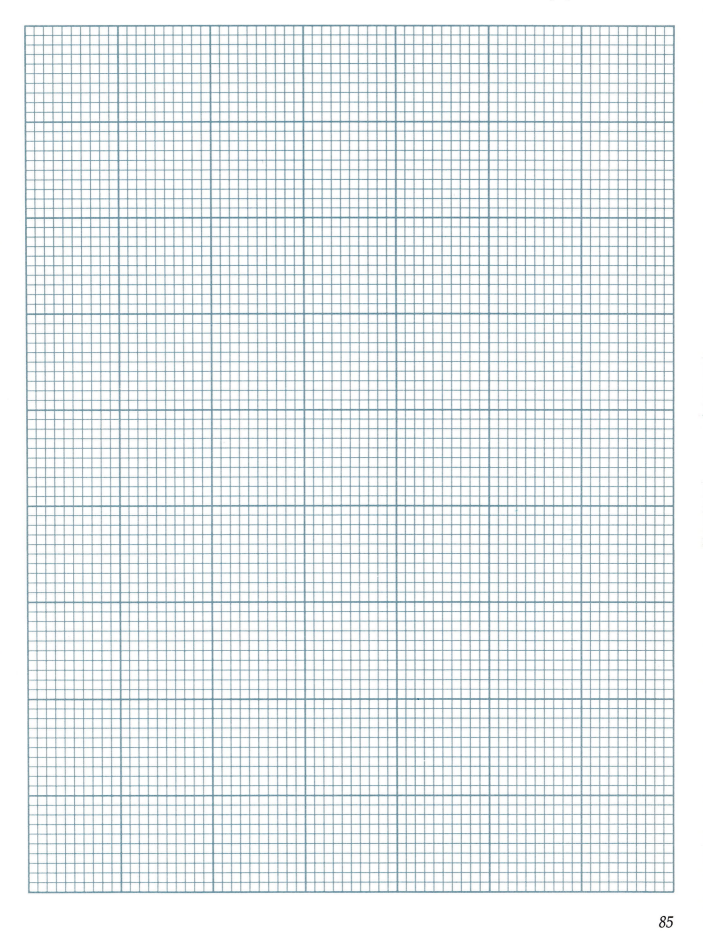

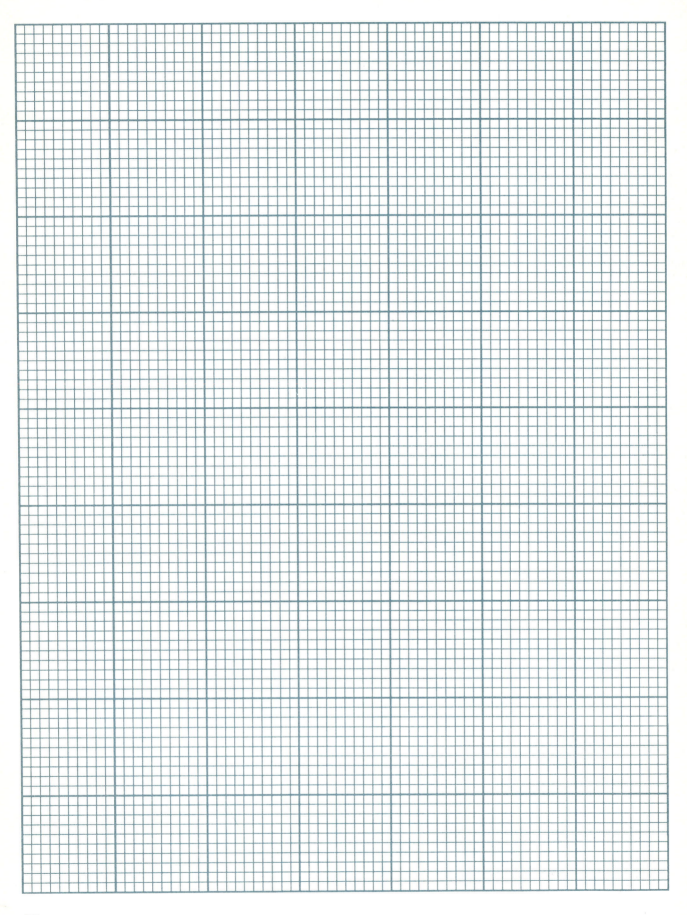

# Electrical Properties of Transporting Epithelia

*Reading assignment: text 28-34*

## Objectives

### Experimental

1. To measure the electrical potential that results from active transport of Na$^+$ across a piece of frog skin.

2. To measure the electrical current corresponding to net movement of positively charged ions across a piece of frog skin.

3. To determine the effect of varying the anion in the extracellular fluid on the electrical characteristics of frog skin.

4. To demonstrate the effects of the metabolic inhibitor 2,4-dinitrophenol and the Na$^+$-K$^+$ pump inhibitor ouabain on active Na$^+$ transport.

### Conceptual

After completing this exercise and the reading assignment, you should be able to:

1. State the direction of Na$^+$ transport across the epithelium of frog skin.

2. Define the terms **homeostasis, active transport, transepithelial potential, anion, cation, short-circuit current,** and **uncoupler.**

3. Explain the origin of the transepithelial potential in Na$^+$-transporting epithelia.

4. Distinguish between **voltage** and **current** and state the units used to measure each.

5. Describe the parts of an Ussing chamber and explain how this apparatus is used to measure the electrical properties of frog skin.

6. Describe the short-circuit method for measuring active transport of ions.

7. Describe the effect of substituting Na$_2$SO$_4$ for NaCl in the extracellular fluid and explain what causes this effect.

8. State the intracellular sites of action of **2,4-dinitrophenol** and **ouabain** and explain their effects on the electrical properties of frog skin.

## Background

Most frogs live in fresh water, which contains about **1-3 mM NaCl**, while the extracellular fluid of a frog's body contains about **110 mM NaCl**. Similarly, the concentration of Cl$^-$ is greater inside a frog than in its environment. In the absence of processes that maintain a stable internal environ-

ment, or **homeostasis**, diffusion of Na$^+$ and Cl$^-$ from the frog into the pond would quickly deplete the frog's extracellular fluid of NaCl and expose its cells to hypotonic stress. In this exercise, we will explore the homeostatic mechanisms that enable frogs to live in a hypotonic environment. Because a frog's environment is hypotonic, we would expect water to move into and Na$^+$ ions to move out of a frog in a fresh water pond, unless these passive processes are counteracted by homeostatic controls.

In **active transport** substances are moved against the driving force of an electrical or a concentration gradient (see Exercise 5). Active transport requires energy other than the energy contained in these driving forces; this energy must be supplied by metabolic reactions within the tissues. To counteract the movement of $Na^+$ and $Cl^-$ out of a frog's body, we would expect to find active transport moving these ions across the frog skin in the opposite direction: from the external surface of the skin to the internal surface.

The homeostatic mechanisms that allow frogs and other vertebrates to survive in fresh water have been studied extensively by comparative physiologists since the early part of this century. In the 1940s a Danish physiologist named Hans Ussing developed methods of studying skin removed from a frog and mounted between two chambers, called **Ussing chambers**. You will use some of his methods in this exercise.

Using two radioactive isotopes of $Na^+$, Ussing showed that when frog skin is bathed on both sides in Ringer's solution, which is similar in composition to the frog's extracellular fluid, there is, as we would expect, a net movement of $Na^+$ from the external side of the frog skin to the internal side. In other words, the unidirectional flux rate of $Na^+$

movement from the external side to the internal side of the skin is larger than the flux rate of $Na^+$ in the opposite direction. This net movement of $Na^+$ occurs in the absence of external driving forces; thus, it must be due to active transport.

Over the next several decades, Ussing and others directed their investigations at the mechanism of $Na^+$ transport by the frog skin. Some questions that arose immediately were:

> ** **Is the active transport carried out by an enzyme?**
>
> ** **Where does the energy for active transport come from?**
>
> ** **How is movement of $Cl^-$ brought about?**

The internal side of a frog's skin is positive relative to the external side. If $Na^+$ moves from the outside to the inside of the frog skin, it must do so against an electrical gradient. **Voltage** (expressed in volts [V]) is a measure of the work required to move electrical charges from one point to another of higher potential. In other words, voltage measures the electrical potential difference between two points. In isolated frog skin bathed in Ringer's solution, the movement of $Na^+$ against an electrical gradient generates a voltage of several tens of mV. This voltage across frog skin is the **transepithelial potential**.

*Figure 6-1. Apparatus used by Ussing.*

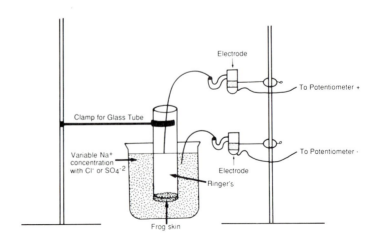

The finding that the inside of frog skin is positive relative to the outside is consistent with Ussing's finding of net $Na^+$ transport into the frog. A transepithelial potential with an inside-positive polarity opposes net $Na^+$ movement towards the blood side. This means that some other force must drive $Na^+$ across the tissue; the movement of $Na^+$ to the inner surface must be due to active transport, powered by cellular metabolism.

The inside-positive transepithelial potential across frog skin could cause net passive movement of other ions in the solution. For example, $Cl^-$ would be expected to move from the external side to the blood side and thus to follow $Na^+$ into the animal. The ease with which $Cl^-$ follows $Na^+$ across the skin (that is, the permeability of the skin to $Cl^-$) affects the magnitude of the transepithelial potential. Since $Cl^-$ follows readily, some of the positive charges from $Na^+$ are canceled, and the transepithelial potential is less than the maximum driving force of the active process. However, if $Cl^-$ did not enter readily, the transepithelial potential would approach the maximum driving force of the active process, and no further net movement of $Na^+$ would be possible.

In this exercise, you will clamp a piece of frog skin between two Ussing chambers (Fig. 6-1), so that you can place various solutions on the inside and the outside surfaces of the epithelium. The solutions are aerated (so that the skin's mitochondria will receive oxygen) by two tubes. You will use the Ussing chambers to measure the electrical properties of frog skin under different conditions.

Measuring the amount of $Na^+$ that is actively transported into a frog is not as straightforward as it might sound. When $Na^+$ is transported inside the frog, the inner surface of the skin becomes more positive. This creates an electrical gradient for $Cl^-$ ions outside the cell. Since the skin is permeable to chloride ions, $Cl^-$ will move passively into the cell, following the electrical gradient, and this will affect your measurements of the voltage caused by inward, active transport of $Na^+$. In other words, because of the movement of negative charges into the cell, the potential difference across

the frog skin cannot be taken as an indication of the magnitude of sodium transport taking place.

One way to get around this problem is to measure **current** instead of voltage. **Current** is the flow of charge. The direction of current flow is the direction moved by a positive charge. (Charge is measured in units of **coulombs** [C]. A current of 1 C/sec = 1 **ampere** [Amp]). To convert current to rate of $Na^+$ movement, see Box.

---

1 Amp = 1 C/sec

1 $\mu$Amp = 1 $\mu$C/sec = $6.241 \times 10^{12}$ electrons/sec

$$\frac{1\ \mu Amp}{1\ \mu Mole} = \frac{6.241 \times 10^{12}\ \text{electrons/sec}}{6.023 \times 10^{17}\ Na^+/\mu Mole}$$

Therefore:

1 $\mu$Amp = $1.036 \times 10^{-5}$ $\mu$Moles $Na^+$/sec

---

A net movement of $Na^+$ should result in an electrical current across the skin, unless there is simultaneous active transport of a negatively-charged ion (**anion**) in the same direction, or another positively-charged ion (**cation**) in the opposite direction. The current generated by active $Na^+$ transport can be measured accurately only in the absence of driving forces. To do this, chemical and electrical driving forces should be abolished.

Ideally, chemical driving forces can be eliminated by bathing the tissue in identical solutions on both sides. However, because the skin is not very permeable to $Na^+$, violation of this condition does not appreciably affect experimental results. Electrical driving forces can be abolished by "clamping" the transepithelial voltage of the tissue at zero. In this procedure the movement of positive charge, caused by active transport of $Na^+$ across the frog skin, is balanced by supplying an equivalent amount of negative charge with electrodes. This allows the voltage across the skin to be kept equal to zero. The current needed to bring the voltage to zero is equal to that generated by active $Na^+$ transport and is called the **short-circuit current**. Any net movement of charge across the skin when it is short-circuited must be due to active transport.

When the tissue is perfectly short-circuited, the current-passing electrode on the blood side matches each $Na^+$ positive charge that appears on that side with a negative charge, and the electrode on the external side compensates each positive charge lost from solution on that side by taking up a negative charge. Experiments with $Na^+$ isotopes have confirmed that, under these conditions, the current, expressed as charges/sec, corresponds closely to the net $Na^+$ movement to the inner side of the skin. Thus,

> ** the short-circuit current is a quantitative measure of active transport taking place across the frog skin.

By measuring how much charge you must supply to keep the voltage equal to zero, you can measure the rate at which charges cross the membrane.

In this exercise, you will perform several experiments designed to provide information about the processes responsible for the movement of $Na^+$ and $Cl^-$ across the epithelium of frog skin. In Part 1, you will measure the potential difference across a piece of frog skin, and you will also measure the short-circuit current across the skin. You will do this using a variety of concentrations of $Na^+$ in the external solution. This will allow you to test the following:

**Hypothesis 1:** The rate of active transport of $Na^+$ depends on the concentration of the substrate, $Na^+$.

If active transport of $Na^+$ depends on enzymes, then $Na^+$ can be regarded as a substrate. If Hypothesis 1 is correct, enzyme activity (that is, the rate of $Na^+$ transport) should increase initially

and should eventually level off as the concentration of $Na^+$ (substrate) increases (see Fig. 3-2 to review this concept).

In addition, in Part 1 some of you will bathe the frog skin in a solution containing $Cl^-$, while others will use an external solution containing $SO_4^{-2}$. Frog skin is impermeable to $SO_4^{-2}$. This experiment will allow you to test another hypothesis:

**Hypothesis 2:** The electrical potential generated by active $Na^+$ transport provides the energy for $Cl^-$ absorption.

Another way of saying this is that the movement of $Cl^-$ into the skin is driven by the inside-positive gradient across the skin. If Hypothesis 2 is correct, then when you bathe the skin in a solution containing an anion (such as $SO_4^{-2}$) which cannot cross the skin, the transepithelial potential should increase because $Na^+$ will be transported to the inside of the skin, but the external anions will be unable to cross the skin.

In Part 2 you will test two additional hypotheses about the mechanism of $Na^+$ transport:

**Hypothesis 3:** Energy for active transport of $Na^+$ is provided by the hydrolysis of ATP.

If Hypothesis 3 is correct, then short-circuit current should decrease in the presence of inhibitors of phosphorylation. You will investigate the importance of metabolic energy for net $Na^+$ transport with a chemical inhibitor of energy metabolism, **2,4-dinitrophenol**. This compound is an **uncoupler**, that is, it abolishes the $H^+$ gradient across the cristae of mitochondria, preventing generation of ATP by oxidative metabolism.

**Hypothesis 4:** Na⁺ transport across epithelial cells is driven by the process responsible for maintaining a low Na⁺ concentration and a high K⁺ concentration in the cytoplasm.

If Hypothesis 4 is true, Na⁺ transport by frog skin should be inhibited by **ouabain**, an inhibitor the Na⁺-K⁺ pump.

## Procedure

Record your results in the laboratory report section at the end of this exercise.

 *Part 1. Effect of varying Na⁺ concentration and external anion on transepithelial potential*

**1. Work in pairs. Each member of the pair should obtain the following (Fig. 6-2):**

> Ussing chamber (<u>handle with care</u>),
> double-pithed frog,
> 3 solutions:
>   Ringer's solution,
>   220 mM sucrose solution,
>   sodium solution, either
>     a) 110 mM NaCl or
>     b) 55 mM Na₂SO₄ + 55 mM sucrose.

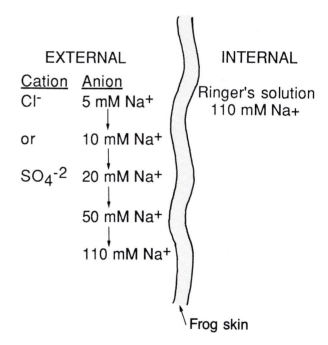

*Figure 6-2. Experimental set-up for studying electrical properties of frog epithelium.*

**The sodium solution and sucrose solutions will be used to make up external solutions with different concentrations of Na⁺. You will do this by**

varying the concentrations of the sodium solution and the sucrose solution.

The reason for adding the 55 mM sucrose to solution (b) is to make both of the sodium solutions the same osmolarity (110 mM). Because the osmolarities are the same, the sucrose solution and the sodium solution can be mixed together to form solutions of varying sodium concentrations without affecting osmolarity.

The reason for having two sodium solutions is to allow one member of the pair to use solutions in which the external anion is $Cl^-$ while the other member uses $SO_4^{-2}$ as the external anion.

2. Prepare the external solutions. One of you should use the NaCl solution and the other should use the $Na_2SO_4$ solution in the external chamber (Fig. 6-2). Make up the solutions with different sodium concentrations for the outer bathing solution, as follows:

| Desired Concentration of $Na^+$ | Sucrose solution (220 mM) | Sodium solution[1] |
|---|---|---|
| 5 mM | 60 ml | 3 ml |
| 10 mM | 50 ml | 15 ml |
| 20 mM | 45 ml | 10 ml |
| 50 mM | 30 ml | 25 ml |
| 110 mM | 0 ml | 50 ml |

[1] Either 110 mM NaCl or 55 mM $Na_2SO_4$ + 55 mM sucrose.

3. Set up the frog skin preparation. Glue one ring onto the frog belly with superglue. Hold it in place for 30 sec. Cut the ring and attached skin from the frog and mount in the Ussing chamber. Make sure that you do not allow the skin to dry out. This will damage the epithelial cells. Use a little grease on the seats for the plastic rings. Combine and mount the chamber.

4. Insert voltage sensing and current delivering electrodes (be careful, the electrodes are very delicate) and fill the two chambers. The inside should be filled with Ringer's. The outside should be filled with 110 mM $Na^+$ solution. (Be sure you use the right solution, either NaCl or

$Na_2SO_4$.) Fix any serious leaks by applying silicone vacuum grease. Accept minor leaks and get a beaker under them. You should see a current of at least 15 μAmp.

5. Vary the sodium concentration of the external solution. If the system is behaving, drain the outside solution and replace it with the 5 mM $Na^+$ solution. If the system is not working well, consult your instructor.

6. After a steady-state has been reached (usually 15 to 25 min), measure the current. Record your results in Table 6-1 (for NaCl solutions) or Table 6-2 (for $Na_2SO_4$ solutions).

7. Drain the solution on the external side, put in the 10 mM $Na^+$ solution, and repeat Step 6.

8. Continue with the 20, the 50, and the 110 mM $Na^+$ solutions. The last solution is the same sodium concentration as Ringer's solution.

 *Part 2. Effect of 2,4-dinitrophenol and ouabain on short-circuit current*

1. When you are through examining the relationship between transport and $Na^+$ concentration, the skin will be in its original condition, with normal Ringer's solution on both sides. One partner should add 1 ml of a solution of 2,4-dinitrophenol to the inside of the skin, while the other should add 1 ml of a solution of ouabain. (CAUTION: Handle the 2,4-dinitrophenol and the ouabain with care. Do not allow it to get on your skin. If it does, immediately wash it off with plenty of water.)

2. Measure the voltage every 5 min for about 30 min.

3. Record your results in Table 6-3 (for 2,4-dinitrophenol) or 6-4 (for ouabain).

4. When you have completed your experiments, carefully disassemble the Ussing chamber and wash it.

# Laboratory Report

## Exercise 6: Electrical Properties of Transporting Epithelia

*Name:* ______________________

*Date:* ______________________

*Lab Section:* ______________________

## Analyzing Your Data

In Part 1, half of you obtained your data using NaCl in the external solution, and half of you used $Na_2SO_4$. Use your data to plot a) potential difference (in mV) and b) short-circuit current (in μAmp) as functions of the concentration of $Na^+$ outside the frog skin. In addition, plot your lab partner's data for the anion you did not use (either $Cl^-$ or $SO_4^{-2}$).

Plot short-circuit current as functions of time after addition of a) 2,4-dinitrophenol and b) ouabain. (Again, share data with your lab partner.)

***Table 6-1.*** *Relationship between concentration of $Na^+$ in external solution and electrical properties of frog skin with $Cl^-$ in external solution.*

| $Na^+$ concentration in external solution | Potential difference (mV) | Short-circuit current (μAmp) |
|---|---|---|
| 5 mM | | |
| 10 mM | | |
| 20 mM | | |
| 50 mM | | |
| 110 mM | | |

**Table 6-2.** *Relationship between concentration of Na$^+$ in external solution and electrical properties of frog skin with SO$_4^{-2}$ in external solution.*

| Na$^+$ concentration in external solution | Potential difference (mV) | Short-circuit current (μAmp) |
|---|---|---|
| 5 mM | | |
| 10 mM | | |
| 20 mM | | |
| 50 mM | | |
| 110 mM | | |

**Table 6-3.** *Electrical properties of frog skin following addition of 2,4-dinitrophenol to internal fluid.*

| Min after addition of 2,4-dinitrophenol | Potential difference (mV) | Short-circuit current (μAmp) |
|---|---|---|
| 5 | | |
| 10 | | |
| 15 | | |
| 20 | | |
| 25 | | |
| 30 | | |

Table 6-4. *Electrical properties of frog skin following addition of ouabain to internal fluid.*

| Min after addition of ouabain | Potential difference (mV) | Short-circuit current (µAmp) |
|---|---|---|
| 5 | | |
| 10 | | |
| 15 | | |
| 20 | | |
| 25 | | |
| 30 | | |

## Questions

1. a) In the absence of active transport, would there be a net movement of $Na^+$ across a frog's skin? _____
Explain.

   b) If so, would this movement be from the inside to the outside or from the outside to the inside of the frog? _____
Explain.

2. Describe a homeostatic mechanism that enables frogs to live in hypotonic environments.

3. Give two characteristics of active transport.

4. What causes the transepithelial potential across a frog's skin?

5. What is the difference between voltage and current?

6. What is a short-circuit current?

7. Which is a better measure of active transport in the frog skin, short-circuit current or potential difference? _____ Explain.

8. Is frog skin permeable to $Cl^-$? _____

9. Is frog skin permeable to $SO_4^{-2}$? _____

10. What happened to transepithelial potential when $SO_4^{-2}$ is substituted for $Cl^-$ in the external solution?

_____

11. What happened to the short-circuit current after 2,4-dinitrophenol was added? _____ Explain.

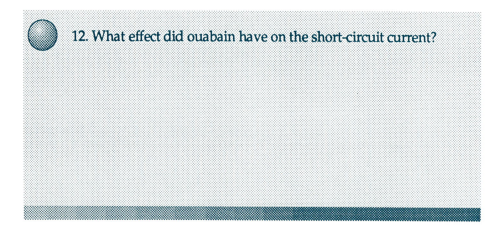

12. What effect did ouabain have on the short-circuit current?

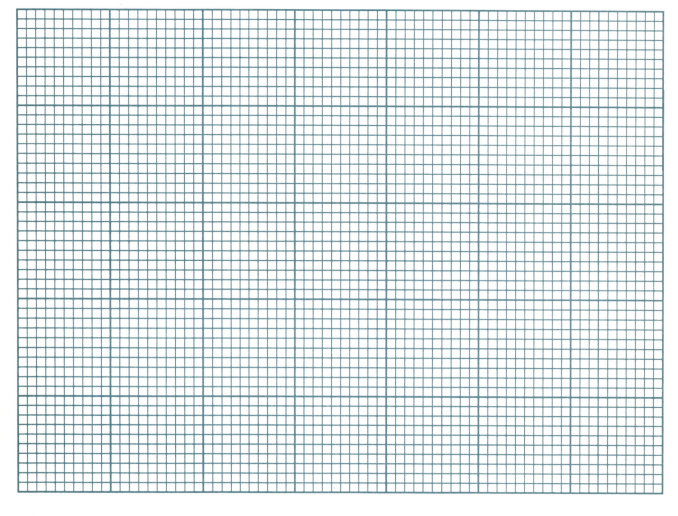

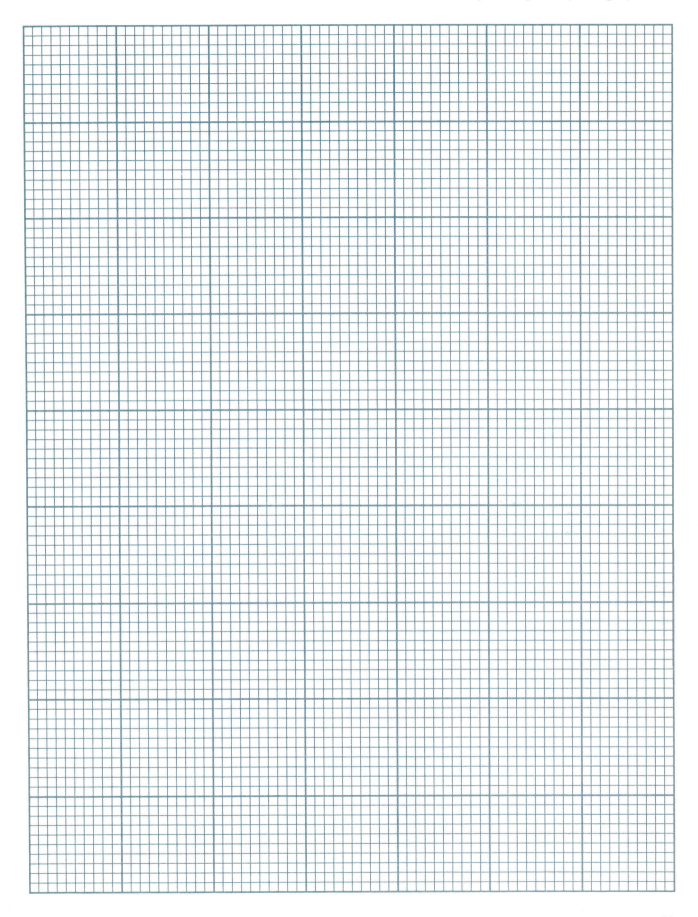

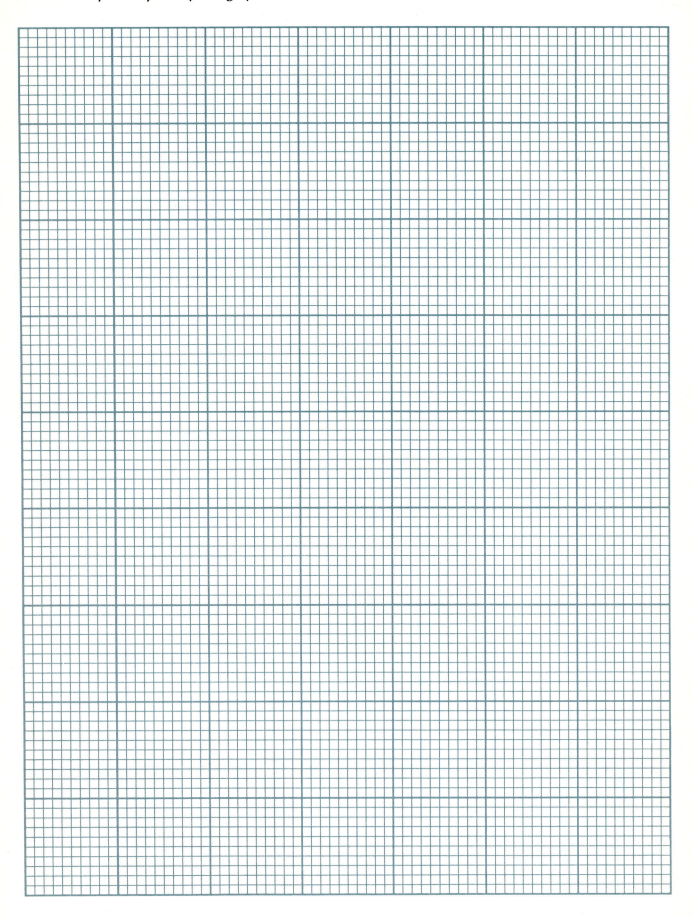

# Compound Action Potential

*Reading assignment: text 139-162*

## Objectives

### Experimental

1. To observe threshold and maximal responses in a compound action potential.

2. To determine the relationship between strength and duration of stimuli that produce a detectable response in the sciatic nerve.

3. To determine the relationship between strength and duration of stimuli that produce a maximal response in the sciatic nerve.

4. To observe separate peaks in a compound action potential representing action potentials in different size classes of neurons.

### Conceptual

After completing this exercise and the reading assignment, you should be able to:

1. Describe the structure of a motor neuron.

2. Define the terms **neuron, dendrite, axon, resting membrane potential, voltage-gated channel, action potential, depolarization, repolarization, propagation, compound action potential,** and **nerve impulse.**

3. Give the resting membrane potential of a typical nerve cell and describe two factors that produce it.

4. Describe the events that produce an action potential.

5. Explain why an action potential in a single axon is an all-or-none response while a compound action potential is a graded phenomenon.

6. Define the term **stimulus artifact** and explain how it is produced.

7. Explain how electrical changes in a nerve produce deflections on an oscilloscope.

8. Describe the relationship between stimulus strength and duration for threshold and maximal responses in a nerve trunk and explain the reasons for this relationship.

9. Describe the relationship between axon diameter and propagation velocity.

10. Explain why the compound action potential has separate peaks that can be observed when the recording electrodes are located several cm from the stimulating electrodes.

## Background

In this exercise, you will again measure electrical phenomena in living tissues. Previously (Exercise 6) you measured electrical events in frog skin; today you will observe the response of the **sciatic nerve** to electrical stimulation. In exercises 8 and 9 you will use stimulation of the sciatic nerve to make muscles contract.

**Figure 7-1.**
*Structure of a motor neuron.*

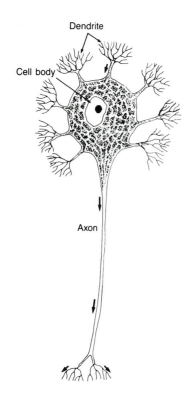

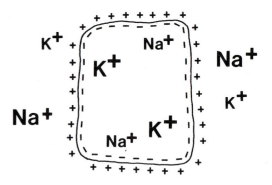

*Figure 7-2. Charges across a cell membrane and concentrations of Na⁺ and K⁺ inside and outside a cell. Sizes of letters represent relative concentrations of sodium and potassium ions. The inner surface of the membranes of neurons and other cells are negatively charged with respect to the outer surface.*

A motor neuron is composed of three regions (Fig. 7-1): a) a **cell body**, b) numerous, relatively short **dendrites**, which conduct impulses toward the cell body, and c) a single, elongated **axon** that conducts impulses away from the cell body toward other neurons or effector organs.

If a microelectrode is inserted into a neuron, the inside of the neuron will be found to be negatively charged with respect to the outside (Fig. 7-2); in other words, neuronal membranes are **polarized**. The **resting membrane potential**, or charge on the inside of a neuron with respect to the outside, is approximately –70 mV. Membrane potentials represent stored energy, and changes in membrane potential serve as the basis for conveying information in excitable cells.

The inside-negative resting membrane potential is brought about primarily by two factors:

a) The concentration gradients of sodium and potassium ions. Sodium ions are more concentrated outside the cell than inside it, and potassium ions are present in greater concentration within the cell than outside it (Fig. 7-2). Active transport in the form of the $Na^+$-$K^+$ pump (see Exercise 6) plays an important role in maintaining these concentration differences.

b) The differential permeability of the cell membrane to $Na^+$ and $K^+$. Membrane permeability is thought to depend on membrane proteins that act as pores through the membrane. Channels are typically selective for particular ions. Each channel may be opened or closed at any given instant. Excitable cells are characterized by the presence of **voltage-gated channels**, which open and close in response to membrane potential.

In addition, the presence within cells of negatively charged proteins that are too large to diffuse through the membrane also contributes to the inside-negative potential. The resting membrane potential of a cell may be altered by changing one of the factors mentioned above: either a change in the permeability to a particular ion or a change in the concentration gradient for an ion.

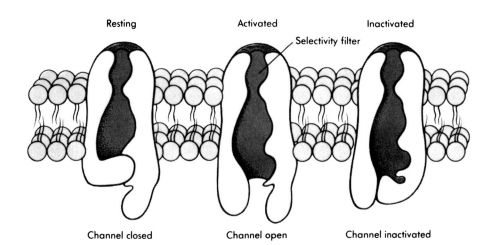

**Figure 7-3.** *A hypothetical model of the Na⁺ channel illustrating the three states: resting, activated and inactivated.*

Stimulation of a neuron causes an increase in membrane permeability to Na⁺ by opening **Na⁺-gated channels** (Fig. 7-3). This allows Na⁺ to passively enter the nerve cell down both its concentration and electrical gradients. The influx of positive charge causes **depolarization**, that is, the charge on the inside of the membrane increases (Fig. 7-4). These events set up a positive feedback cycle in which increases in membrane permeability to Na⁺ allow more Na⁺ to move down its concentration and electrical gradients into the cell, and this net addition of positive charge to the inside of the membrane causes further depolarization (Fig. 7-5).

Once a Na⁺-gated channel has opened, it closes spontaneously after a time. Depolarization also opens K⁺ channels, which open more slowly than Na⁺ channels, and the outward flow of K⁺ along its concentration gradient contributes to the reestablishment of the resting membrane potential (**repolarization**).

The rapid reversal and reestablishment of membrane potential is called an **action potential** (Fig. 7-4). The magnitude of an action potential is constant; it is not related to the size of the stimulus. Therefore, the action potentials of individual neurons are **all-or-none** events.

Transient, weak stimuli do not initiate the positive feedback cycle leading to an action potential. Only stimuli that depolarize the membrane to a **threshold potential** initiate action potentials. Threshold

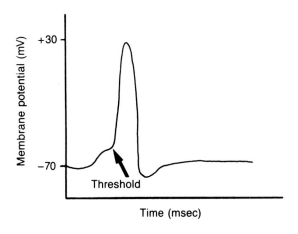

**Figure 7-4.** *Changes in membrane potential of a neuron during an action potential.*

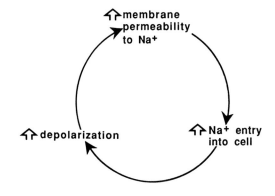

**Figure 7-5.** *Positive feedback relationship between membrane permeability to Na⁺ and depolarization during an action potential.*

is defined as the minimum value of a membrane potential at which an action potential will occur 50% of the time (Fig. 7-4).

An action potential is a mechanism used for long-distance signaling in nerves and muscles. Action potentials are capable of **propagation**; that is, once initiated an action potential can spread along an axon. This occurs because the $Na^+$ ions that cross the membrane upset the balance between positive and negative charges in the interior of the cell. The entry of $Na^+$ depolarizes adjacent regions of the cell membrane, causing action potentials to travel along the membrane as adjacent regions are brought to threshold (Fig. 7-6).

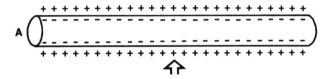

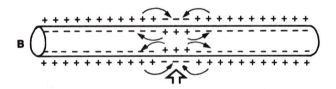

*Figure 7-6. Propagation of an action potential along a membrane of a neuron. Open arrow indicates site of stimulation (A). Positive charges move toward negative charges, causing depolarization to spread in both directions away from site of action potential (B).*

The **nerves** that you can see when you dissect a frog are composed of the axons of many neurons. For instance, the sciatic nerve (Fig. 7-7) consists of hundreds of axons, which innervate all the muscles in the frog's lower leg and foot. This nerve also contains axons that carry sensory information from the skin and muscle receptors to the spinal cord. Thus, under normal circumstances (that is, when the nerve is in the frog), the sciatic nerve carries information travelling in two directions: some axons carry information toward the central nervous system from the sensory receptors and others

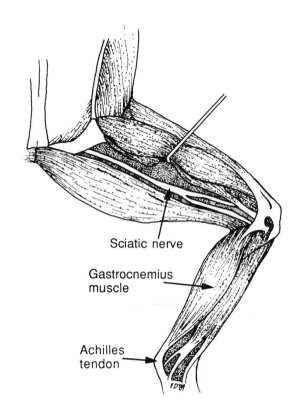

*Figure 7-7. Hind limb of a frog, showing location of the sciatic nerve.*

conduct impulses away from the central nervous system to the muscles. When you stimulate a nerve trunk such as the sciatic, you will observe a **compound action potential**, that is, a composite of the responses of many individual axons. The collective activity of the axons in the nerve is known as the nerve impulse.

The electrical activity associated with nerve impulses can be conveniently observed with an **oscilloscope**. This instrument uses a beam of electrons emitted from a cathode. The electron beam is driven from left to right across a phosphorescent screen. When the beam hits the screen it emits a dot of light (Fig. 7-8B). Because vertical deflection of the beam is proportional to voltage, the image on the screen represents a plot of voltage, on the $y$ axis, as a function of time, on the $x$ axis.

In this exercise, the sciatic nerve will be removed from a frog, and laid across several wires in a

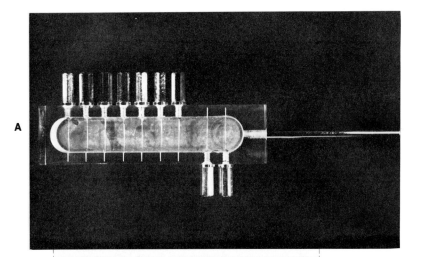

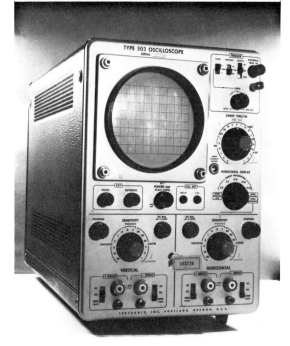

*Figure 7-8.* (A) *Nerve box.*
(B) *Oscilloscope.*

produces a increase in **amplitude** (height) of the stimulus artifact.

When the stimulating electrodes deliver a pulse of electricity to the nerve, action potentials will travel along the nerve in those axons that are depolarized to threshold. The nerve impulse will reach the first recording electrode (Fig. 7-9: wire 1) before it reaches the second recording electrode (Fig. 7-9: wire 2). A potential difference between the two recording electrodes produces a vertical deflection of the electron beam (Fig. 7-11).

The following events occur as impulses travel down the nerve:

1. Initially, when there is no impulse in the region of the nerve between the wires, there is no electrical difference between the two wires. In this case, there is no deflection of the signal on the oscilloscope (Fig. 7-11A).

2. When a nerve impulse travels down the nerve from the point of stimulation, the signal reaches wire 1 first. This point becomes depolarized relative to the unchanged region of the nerve recorded by wire 2. Depolarization is seen on the screen as a positive, or upward, deflection (Fig. 7-11B).

**nerve box** (Fig 7-8A). A pair of **stimulating electrodes** from a **stimulator** will deliver a pulse of electricity to one point on the nerve. The nerve will also be in contact with a pair of **recording electrodes** connected to the cathode-ray tube of the oscilloscope (Fig. 7-9). The delivery of a pulse of electricity from the stimulus electrodes results in a small vertical deflection at the beginning of the horizontal sweep of the electron beam (Fig. 7-10). This deflection is the **stimulus artifact**. An increase in stimulus strength

**Figure 7-9.** *Use of oscilloscope to observe electrical activity in a nerve. Arrows indicate direction of electrical impulse.*

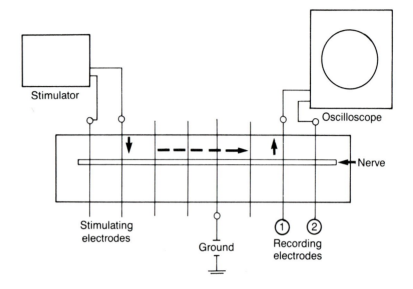

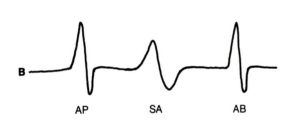

**Figure 7-10.** *Tracing from an oscilloscope. At low voltage only stimulus artifact, SA, is visible (A). At threshold voltage, action potential, AP, and stimulus artifact are seen (B).*

3. As the impulse continues moving down the nerve, it passes the region of wire 1. When it is between wire 1 and wire 2, there is again no difference in the electrical signals recorded by the two wires, and the deflection on the oscilloscope returns to the baseline indicating zero potential difference between the wires (Fig. 7-11C).

4. Next, the impulse arrives at the region of the nerve recorded by wire 2, and that region becomes depolarized. However, because the oscilloscope always shows the potential difference recorded by wire 1 relative to wire 2, the signal recorded by wire 1 is now negative relative to the depolarized signal recorded by wire 2. This appears as a downward deflection on the oscilloscope (Fig. 7-11D).

5. Once the impulse has passed wire 2, there is again no potential difference between wire 1 and wire 2, and the deflection on the oscilloscope returns to zero (Fig. 7-11E).

Because of the way in which electrical current spreads across membranes, larger axons reach threshold potential in response to relatively small currents. Consequently, the axons of the sciatic nerve differ in threshold; some axons will be brought to threshold by weaker electrical stimuli than others. Increasing stimulus voltage causes more axons to be brought to threshold, and the size of the compound action potential decreases. Thus, the compound action potential is a **graded phenomenon**; it is not all-or-none (even though the action potentials of its constituent axons are).

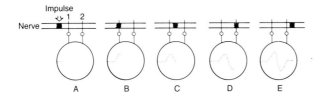

*Figure 7-11A-E. Electrical events in an axon during an action potential and resulting tracing seen on oscilloscope. Numbers indicate wire 1 and wire 2, the two recording electrodes. See "Background" for details.*

The lowest voltage that produces a detectable response in the nerve is the **threshold stimulus** (Fig. 7-10), and the response is a threshold response. (Note that **threshold potential** describes the membrane potential required to produce a stimulus, while **threshold stimulus** describes a voltage that is applied to the nerve and produces a response that can be detected with the aid of the oscilloscope.) As stimulus voltage is increased, the amplitude of the deflection produced on the oscilloscope screen also increases, until a point is reached at which no further

increase in amplitude occurs. The stimulus that produces this **maximal response** is the **maximal stimulus** for the nerve.

The stimulator allows you to vary three aspects of stimulus intensity: **frequency, strength,** and **duration** . In Part 2 of this exercise you will keep stimulus frequency and duration constant while varying stimulus strength. This will enable you to determine the threshold and maximal stimulus voltages at a stimulus frequency of 120/sec and a duration of 0.1 msec.

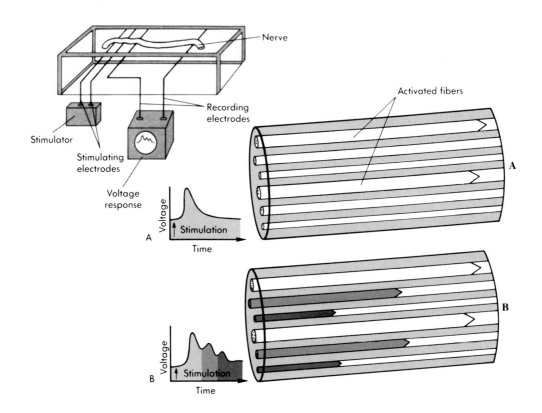

**Figure 7-12.** *Stimulation of a whole nerve causes a compound action potential. Small stimulating currents activate only the largest axons. These have similar conduction velocities and produce a single peak at the recording electrode (A). Larger stimulating currents can activate all classes of axons in the nerve, resulting in multiple peaks as the action potentials of different axon classes arrive at the recording electrode at different times.*

Sodium channels can be opened by brief, intense stimuli or long-lasting stimuli of moderate strength. Consequently, the threshold and maximal voltages depend on stimulus duration. In Part 3 you will investigate this relationship by determining threshold and maximal voltages for stimuli of varying durations.

Large-diameter axons will propagate action potentials more rapidly than axons with small diameters. Within a nerve there are several classes of axons of similar diameter; all axons in one of these groups have similar conduction velocities.

When you produce a maximal response in the nerve, you will not be able to detect different size classes of axons if the recording electrodes are placed near the stimulating electrodes. This is because, in the short time it takes the nerve impulses to reach the recording electrodes, the action potentials traveling in the large, rapidly propagating axons will not be very far ahead of the action potentials moving down the smaller-diameter axons. (Similarly, runners of different abilities will be farther apart at the end of a race than near the beginning.)

However, if you place the recording electrodes several cm from the stimulating electrodes, you should be able to detect more than one peak in the deflection on the oscilloscope (Fig. 7-12). The separate peaks are produced by action potentials in different size classes of axons arriving at the recording electrodes at different times. In Part 3 you will use this technique to identify different classes of axons.

## Procedure

Record your results in the laboratory report section at the end of this exercise.

 *Part 1. Setting up the apparatus*

1. Observe the dissection of the sciatic nerve from a frog that has been double-pithed by your instructor. You will do this dissection yourself in the next exercise.

2. Trace the wire leads from the stimulator to the nerve chamber and from the nerve chamber to the oscilloscope to be sure you understand the pathways.

3. Familiarize yourself with the dials and knobs on the stimulator (Fig. 7-11). The knobs change the characteristics of the stimulus by factors of 10, as follows:

> The voltage knob controls the magnitude of the electrical shock.
>
> The duration knob determines how long the stimulus will be applied.
>
> The frequency knob controls the interval between shocks when you deliver a continuous train of shocks.

A wire connecting the stimulator to the oscilloscope synchronizes their signals, so that the oscilloscope line sweeps across the screen each time the stimulator delivers a stimulus to the nerve. The dials on the oscilloscope control the speed at which the electron beam travels across the screen of the oscilloscope (sweep speed) and the amplitude or degree of magnification of the signal (sensitivity). Your instructor will show you how to set the sweep speed and sensitivity.

**Part 2. Determination of threshold and maximal stimulus intensity at 0.1 msec duration and 120 pulses/sec frequency**

1. Set the stimulator (voltage) knob at its lowest value. Set the duration at 0.1 msec and set the frequency at 120/sec.

2. Gradually increase the voltage, until you see evidence of a response from the nerve on the oscilloscope screen. Record this threshold value in Table 7-1. (Remember that the first deflection on the screen, the farthest to the left, is the stimulus artifact: Fig. 7-9).

3. Continue to increase the voltage gradually until an increase in voltage does not produce an increase in magnitude of the response. Record the voltage that produces this maximal response in Table 7-2 .

 **Part 3. Relationship between stimulus strength and duration**

1. Repeat Part 1: steps 2 and 3 using a duration of 0.3 msec. Record your data in tables 7-1 (threshold response) and 7-2 (maximal response).

2. Repeat Part 1: steps 2 and 3 two more times, increasing the stimulus duration by 0.2 msec each time. When you have completed this part of the exercise, you will have found the threshold and maximal voltages at each of the following durations: 0.1 msec, 0.3 msec, 0.5 msec, and 0.7 msec. Enter your results in tables 7-1 and 7-2.

 **Part 4. Separation of the compound action potential into its components**

1. Using a pair of fine forceps, pinch the nerve firmly at the location of the second stimulating electrode (the one that is farthest from the stimulating electrodes). Check with your instructor before going on to the next step.

2. Reduce the stimulus frequency to 5/sec. Set the duration knob at 0.1 msec and set the stimulator at a value that will deliver a maximal response at that duration. (Determine this value from Table 7-2 and enter it in figs. 7-A and 7-B.)

3. Place the stimulating electrode about 5 mm from the recording electrode.

4. Stimulate the nerve. You should see a single peak on the oscilloscope screen.

5. Sketch your results in Fig. 7-A.

6. Place the stimulating electrode approximately 50 mm from the recording electrode.

7. Stimulate the nerve. You should see more than one peak on the oscilloscope screen.

8. Sketch your results in Fig. 7-B.

# Laboratory Report

**Exercise 7: Compound Action Potential**

*Name:* ______________________________

*Date:* ______________________________

*Lab Section:* ______________________________

## Analyzing Your Data

Use your data from Table 7-1 to plot the relationship between duration (independent variable) and strength (dependent variable) of the stimuli that produced threshold responses in the sciatic nerve.

Do the same with the data from Table 7-2 to show the relationship between stimulus strength and duration for maximal responses.

***Figure 7-B.** Results of stimulation of the sciatic nerve with recording electrode 50 mm from stimulating electrode (stimulus duration: 0.1 msec; stimulus intensity: ________V).*

***Figure 7-A.** Results of stimulation of the sciatic nerve with recording electrode 5 mm from stimulating electrode (stimulus duration: 0.1 msec; stimulus intensity: ________V).*

*Table 7-1. Strength and duration of stimuli producing threshold response in the sciatic nerve.*

| Duration of stimulus (msec) | Strength of stimulus (V) |
| --- | --- |
| 0.1 | |
| 0.3 | |
| 0.5 | |
| 0.7 | |

*Table 7-2. Strength and duration of stimuli producing maximal response in the sciatic nerve.*

| Duration of stimulus (msec) | Strength of stimulus (V) |
| --- | --- |
| 0.1 | |
| 0.3 | |
| 0.5 | |
| 0.7 | |

# *Questions*

1. What are the three parts of a neuron?

2. What is the approximate value (in mV) of the resting membrane potential of a typical nerve cell? _____

3. What is an action potential?

4. What is a compound action potential?

5. a) Sketch the shape of the deflection that is produced on the oscilloscope screen when a nerve is stimulated. (Assume that the nerve has not been crushed.)

   b) Describe the electrical events in the nerve that produce this deflection .

6. What is the major difference between an action potential in a single neuron and a compound action potential?

7. a) Did the amplitude of the response recorded in the sciatic nerve increase in response to an increase in the voltage of the stimulus?

_____

   b) Given that the response of neurons to electrical stimulation is all-or-none, how can you explain these results?

8. a) What is the shape of the curve depicting threshold stimulus voltage as a function of stimulus duration? _____

   b) What causes this relationship between stimulus voltage and duration?

9. Why do impulses travel more rapidly along some axons than along others?

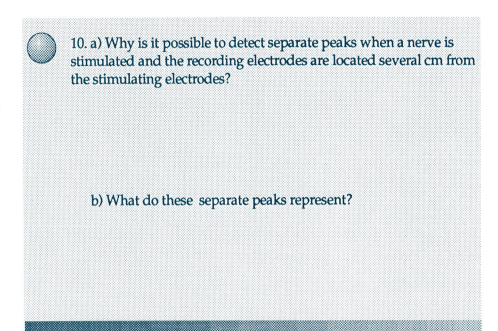

10. a) Why is it possible to detect separate peaks when a nerve is stimulated and the recording electrodes are located several cm from the stimulating electrodes?

b) What do these separate peaks represent?

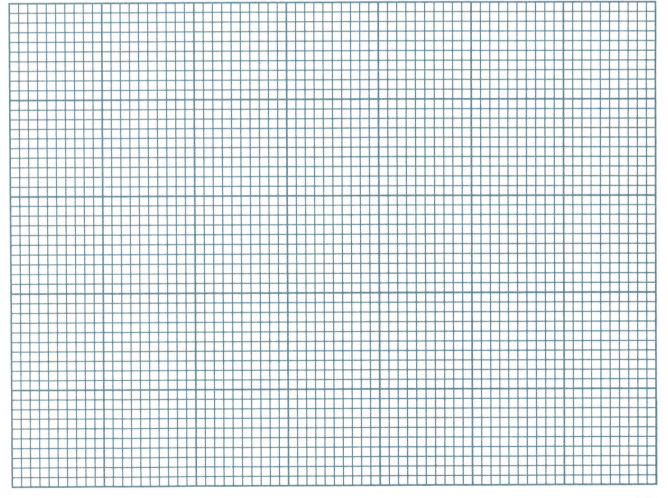

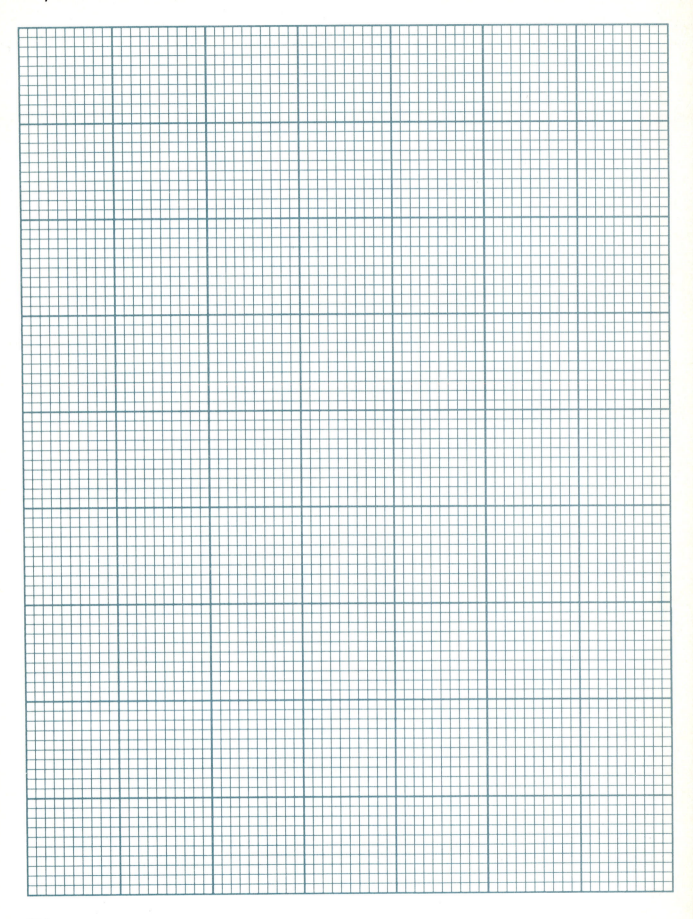

# Motor Nerves and Skeletal Muscle

*Reading assignment: text 292-297*

## Objectives

### Experimental

1. To observe the contraction of the frog gastrocnemius muscle in response to
a) electrical stimulation of the motor nerve and
b) direct electrical stimulation of the muscle.

2. To determine the relationship between stimulus frequency and muscle activity.

### Conceptual

After completing this exercise and the reading assignment, you should be able to:

1. Define the terms **excitability, isotonic, preloaded muscle, neuromuscular junction, twitch, threshold stimulus, optimal stimulus, summation, tetanus,** and **muscle fatigue.**

2. Explain the roles of the following pieces of equipment in the study of isotonic muscle contractions: **stimulator, muscle lever,** and **chart recorder**.

3. Explain why the stimulus required to activate a muscle when its motor nerve is stimulated is much lower than the stimulus required when the muscle is stimulated directly.

4. Explain how multiple stimuli can produce summation of contraction in a skeletal muscle.

## Background

Muscle and nerve cells are **excitable**, that is, they respond to the application of external stimuli by generating action potentials. In muscle cells, action potentials result in contraction. This exercise will make use of the frog **gastrocnemius** (calf) **muscle-sciatic nerve** preparation. In the frog's motor program for a jump, activation of the gastrocnemius occurs along with activation of the muscles of the thigh that rotate the femur and extend the tibio-fibula. In this exercise you will observe the contraction of the muscle in response to single and repeated electrical stimuli applied to the motor nerve and to the muscle itself.

**Tension** is the force a contracting muscle exerts on an object, while **load** is the force exerted on a muscle by an object. In this exercise the muscle will be free to shorten, lifting a lever to which a load is attached and which moves a recording pen (Fig. 8-1). Such contractions are called **preloaded isotonic contractions**. **Preloaded** indicates that the muscle is placed under tension by a load before it is stimulated. **Isotonic** indicates that the tension of the muscle is constant throughout the

117

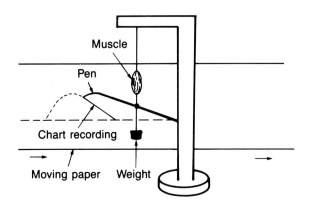

*Figure 8-1.* *General method of recording isotonic contractions in skeletal muscle.*

To measure shortening in the sciatic-gastrocnemius preparation, you will mount the nerve and muscle so that they can be kept moist and are accessible to stimulation (Fig. 8-2). Electrical stimuli will be delivered by a **stimulator**. The muscle will be connected to a **lever**. Movement of the lever causes deflection of a pen on a **chart recorder** (Fig. 8-3). The chart recorder allows a strip of paper to be moved at a controlled rate. In the chart record of isotonic contractions (Fig. 8-1), the horizontal axis is in units of time, and the vertical axis units are the relative distance the load is moved as the muscle contracts and then relaxes.

contraction but the length of the muscle is not constant. In other words, during an isotonic contraction the stimulated muscle is allowed to shorten, lifting the load.

In the intact animal, action potentials travel from motor neurons in the spinal cord along sciatic nerve axons into the gastrocnemius muscle where they result in release of the neurotransmitter **acetylcholine** at the **neuromuscular junctions** (synapses between nerve endings and the muscle) (Fig. 8-4). Each axon branches within the muscle to make synapses on a number of muscle cells. A

*Figure 8-2.* *Diagram of muscle holder and recording apparatus*

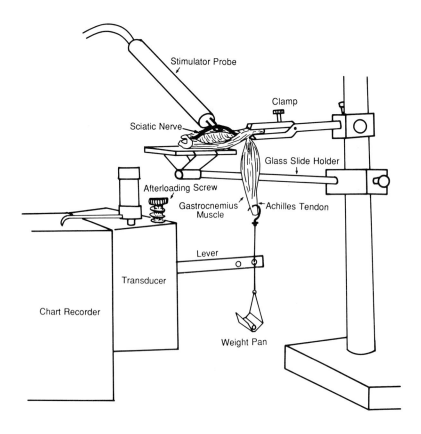

single action potential in a motor nerve causes the release of enough acetylcholine to cause an action potential in each of the muscle cells innervated by that nerve.

The mechanical response in a muscle that results from a single action potential is called a **twitch**. An action potential in the muscle causes release of $Ca^{+2}$ from intracellular stores in the **sarcoplasmic reticulum** (Fig. 8-5). The $Ca^{+2}$ activates the contractile machinery, resulting in a twitch. The duration of a twitch is determined by the rate of reuptake of $Ca^{+2}$ into the sarcoplasmic reticulum. When the $Ca^{+2}$ levels of the cytoplasm have returned to their prestimulus levels, the muscle relaxes.

The lowest stimulus intensity that triggers a twitch is the **threshold stimulus**. As stimulus intensity is increased, more motor axons are recruited, and twitch magnitude increases, until a point is

*Figure 8-3. Chart recorder.*

reached at which further increases in intensity do not produce any further increase in magnitude. The lowest stimulus intensity that evokes a maximal twitch is the **optimal stimulus**. You

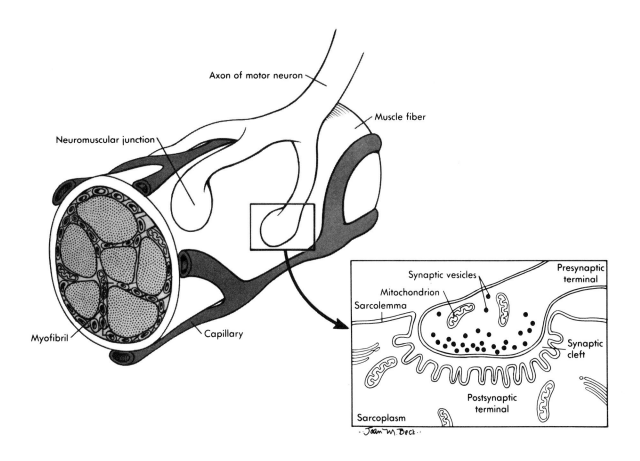

*Figure 8-4. Neuromuscular junction in skeletal muscle.*

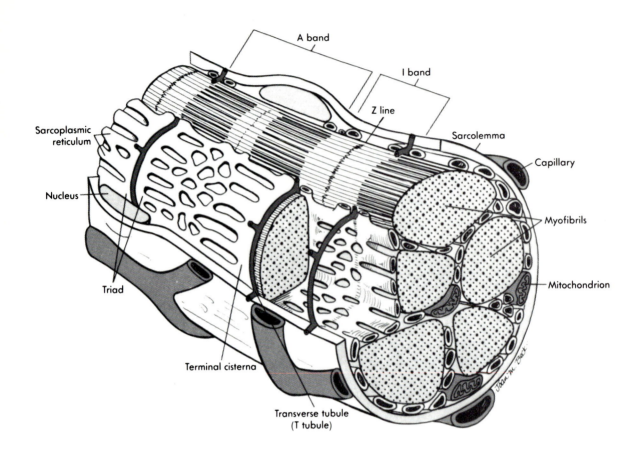

**Figure 8-5.** *Structure of skeletal muscle, showing sarcoplasmic reticulum where $Ca^{+2}$ is stored.*

will determine threshold and stimulus intensities for nerve stimulation and direct stimulation of the muscle in parts 2 and 3 of this exercise.

Twitch duration varies characteristically between muscle types and is also affected by temperature. For the muscle used in this exercise, a twitch should last between 50 and 100 msec at room temperature. This time is insufficient for the muscle to undergo maximum shortening even if it is lifting a very light load.

The duration of the action potential that initiates a twitch is much shorter than the twitch itself. Therefore, if stimulation is repeated at intervals shorter than the twitch duration, the muscle does not fully relax between stimuli, and the resulting peak tension is greater than the tension developed during a single twitch (Fig. 8-6B and C). This response of skeletal muscle to repeated activation of a motor neuron is known as **summation**. High stimulus frequencies result in a smooth, sustained maximal contraction called a **tetanus** (Fig. 8-6D).

If you know the duration of a single twitch, you should be able to predict the minimum frequency at which repeated stimuli will begin to cause summation. Similarly, if you know the time between the onset and the peak of shortening, could you predict the lowest frequency of stimulation that would cause tetanus? In Part 4 you will formulate predictions about stimulus frequencies and then gather data to test them.

Eventually, after prolonged stimulation, a muscle will not be able to sustain continuous contraction. This decrease in the maximal force a muscle can develop and in its ability to sustain a continuous contraction with continued activation is termed **fatigue**. You will demonstrate this phenomenon in Part 5.

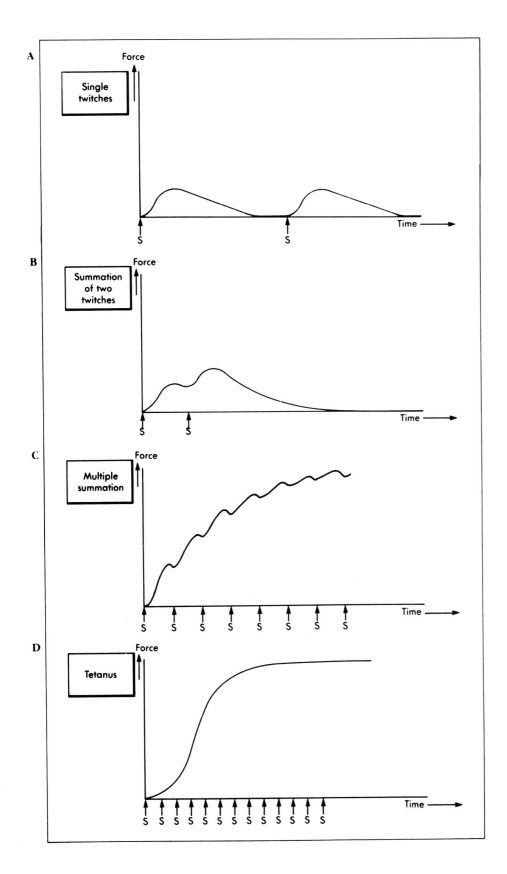

## *Procedure*

Record your results in the laboratory report section at the end of this exercise.

 *Part 1. Dissecting and mounting the nerve and muscle*

1. Arrange the muscle holder and recording apparatus as shown in Fig. 8-2. Notice that moving the lever causes the recording pen to move, and make sure the lever swings freely. Find and identify the stimulator probe and make sure that it is properly connected to the stimulator.

2. Be sure you know how to advance the chart paper and check that the recording pen is making a legible trace. The pens in the ink-writing apparatus are not supposed to become clogged, but ink left in them will dry and clog them. Therefore, they must be left with all the ink returned to the plastic bottles. To start the gravity flow in the pens, gently squeeze the plastic bottle and then, while you are still squeezing, loosen the cap. The pen should write. Squeezing the bottle and then tightening the cap and releasing the pressure should cause the ink to be drawn back into the bottle.

You will need a baseline to compare the heights of different contractions of the muscle. Keep a complete record on the chart recording of everything you do. <u>Each diagram should be clearly labeled and should specify recorder speed and stimulus strength, duration, and frequency</u>. You can write directly on the chart with a felt-tip pen.

3. When you are familiar with the recording apparatus and understand how the living tissue will be mounted, proceed with the dissection. Obtain a double-pithed frog from your instructor. Skin the frog's legs. This is done by cutting the skin carefully around the waist of the animal and peeling it off as one does a pair of panty hose. After you have removed the skin you must keep the exposed tissues moist with Ringer's solution.

4. Part the two major muscles of the frog's thigh with your fingers and find the sciatic nerve. Do not cut the nerve. Using a glass probe, free a length of nerve. Excessive stretching or drying out will damage the nerve, so be careful. Free enough nerve for the two wires of the stimulating probe to be passed under the free part. Lift up the nerve with the probe so that the wires do not touch underlying muscle and hold the nerve and probe in this position.

5. Set the stimulator for multiple stimuli at an intensity of 1 V, frequency of 5 stimuli/sec, and duration of 0.1 msec/stimulus. Stimulate the sciatic nerve. The gastrocnemius muscle should contract. If nothing happens, increase the voltage. Notice that the result is a brisk extension of the frog's foot.

6. Stimulate the muscle directly by touching it with both wires of the stimulator probe. For this, it will probably be necessary to increase the stimulus intensity by several volts. You may also want to try stimulating other muscles of the leg and foot directly to observe their effect on the skeleton.

7. After you have observed the role of the gastrocnemius muscle in movement of the frog, prepare the muscle for mounting in the muscle holder. <u>During this process, remember to keep the exposed tissues moist with Ringer's solution and avoid bruising the sciatic nerve</u>. Insert your blunt probe between the gastrocnemius muscle and the tibio-fibula bone of the lower leg and move the probe toward the frog's foot to loosen the Achilles tendon from the heel. Cut the tendon at the heel, leaving as much tendon as possible attached to the gastrocnemius.

8. Cut away the other muscles that are attached to the tibio-fibula. Then cut through the tibio-fibula itself about two-thirds of the distance from the knee to the ankle, detaching the foot and ankle from the leg. Use your heavy scissors to sever the thigh from the body by cutting the musculature and femur, cutting as close to the body as possible. Use a blunt probe to free the sciatic

nerve, then tuck it back under the surrounding muscle to keep it moist.

9. Now mount the muscle in the holder. Fasten the tibio-fibula in the clamp so that the gastrocnemius hangs down and the sciatic nerve is accessible for stimulation with the probe (Fig. 8-2). Make sure the clamp grips the bone firmly. Attach the Achilles tendon to the recording lever by inserting a hook through the tendon and connecting the hook to the lever with fine wire.

10. When the muscle is connected to the lever, lift the lever manually as far as it will go. Note the position of the recording pen when you do this. If the pen is driven this far by a muscle contraction, an erroneous measure of the extent of shortening will result. Press down on the lever gently to be sure that with the muscle in place the lever is not resting at the bottom of its range. It may be necessary to move the muscle holder up or down slightly to position the recording pen near, but not at, the bottom of its range. Drench the preparation with Ringer's solution.

 *Part 2. Stimulating the motor nerve*

1. Set the stimulator to deliver single pulses of 0.1 msec duration. Set the intensity as low as possible. Make sure the afterloading screw is in the unscrewed position (up). Set the paper speed at 2 cm/sec. It is easiest if one student triggers the stimulator while another raises the nerve with the stimulator probe. Carefully lift the nerve with the probe so that both wires are touching the nerve and neither wire is touching anything else. Deliver a stimulus.

2. If nothing happens, increase the stimulus intensity in steps of 0.1 V. At some point you will see the muscle twitch, and the recording pen will deflect.

3. Keep varying the stimulus intensity until you have reached the threshold value. Record this value in Table 8-1.

4. Continue to make stepwise increases in stimulus intensity. Record a twitch at each intensity, label the trace, and advance the chart record by about 1 cm before recording the next response. (Make sure that the recording pen is not being driven to the top of its range). Wet the muscle and nerve with Ringer's solution and allow the preparation to rest for 5 min.

5. Determine the magnitude of the twitch produced at each stimulus intensity and record these values in Table 8-2. Continue this procedure until further increases in stimulation fail to produce an increase in twitch magnitude.

 *Part 3. Direct stimulation of the muscle*

1. Tuck the sciatic nerve back under the surrounding thigh muscles. Reset the stimulus intensity to the lowest setting. Touch the gastrocnemius muscle with the probe and trigger a single stimulus.

2. Increase the stimulus intensity to find the threshold stimulus intensity for a twitch in response to direct stimulation. Record this value in Table 8-1. Wet the preparation and allow it to rest for 5 min.

3. Repeat Part 2, Step 5. Enter your results in Table 8-2. Wet the preparation and allow it to rest for 5 min.

 *Part 4. Multiple stimulation and summation*

1. If the nerve performed satisfactorily in Part 2, continue using nerve stimulation. If it did not produce satisfactory results, use direct stimulation of the muscle. Set the stimulus intensity at a value slightly in excess of the optimum value to ensure that a maximum response will always be produced. Wet the preparation with Ringer's solution after each bout of stimulation.

2. Set the chart recorder to its maximum paper speed (cm/sec) but do not start the paper moving until you are ready to stimulate. When all is ready, start the paper moving and deliver the stimulus. To avoid waste, stop the paper as soon as you have triggered a response. (If the recorder you are using is not capable of moving the paper at a speed of at least 5 cm/sec, it may not be possible to measure twitch duration accurately. In that case, it will be necessary for you to estimate twitch duration.)

3. Using a metric ruler or the paper scale itself, measure the horizontal distance from the onset of contraction to the peak of shortening and from the beginning to the end of the twitch. Enter these values in Table 8-3.

4. Convert these distances to time as follows:

$$\frac{\text{duration of}}{\text{twitch (sec)}} = \frac{\text{distance (cm) x 1 sec/number}}{\text{of cm traveled in 1 sec.}}$$

Note that the expression 1 sec/number of cm travelled in 1 sec is the inverse of paper speed. When you set the equation up in this way, the cm in the numerator and the denominator cancel out:

$$\text{cm x sec/cm = sec.}$$

This gives you the value for time in sec. You now know the time to peak shortening and the twitch duration. Enter these values in Table 8-3.

5. Using the values obtained in steps 2 to 4, above, predict the minimum frequency required to produce summation and the minimum frequency to produce tetanus. Enter your predicted values in Table 8-3.

6. Set the frequency at 2 stimuli/sec. Set the recorder speed to 0.5 cm/sec. Start the paper moving and stimulate for about 2 sec.

7. Reset the frequency to 5/sec and stimulate again for 2 sec. Repeat at frequencies of 20/sec, 60/sec, and 200/sec. Make sure that the recording pen does not reach the top of its range when you stimulate at higher frequencies. Describe your results in Table 8-3.

8. Use your results obtained in steps 6 and 7 to determine the actual values for the minimum stimulus frequency required to produce summation and minimum stimulus frequency to produce tetanus. Enter these values in Table 8-3.

9. Compare your predicted and actual values for minimum stimulus frequency.

 *Part 5. Fatigue*

1. Set the stimulator for multiple stimuli at the same intensity used in Part 4 and a frequency that produces tetanus in the muscle.

2. Stimulate the nerve continuously for 1 min. Describe your results in Table 8-4.

3. Drench the preparation with Ringer's solution and allow it to rest for 5 min.

4. Stimulate again for 0.5 min. Describe your results in Table 8-4.

# Laboratory Report

## Exercise 8: Motor Nerves and Skeletal Muscle

*Name:*

*Date:*

*Lab Section:*

## Analyzing Your Data

Your chart recording is the primary record of data for this exercise. Use it to determine the values required to fill in tables 8-1 to 8-4. Then answer the questions that follow.

***Table 8-1.*** *Stimulus intensity producing threshold response in the frog gastrocnemius.*

| Stimulus duration: | Stimulus frequency: |
|---|---|
| **Type of stimulation** | **Threshold stimulus intensity** |
| Nerve | |
| Muscle | |

**Table 8-2.** *Relationship between stimulus intensity and twitch magnitude in the frog gastrocnemius.*

| Stimulus duration: | Stimulus frequency: |
|---|---|
| **Stimulus intensity** | **Twitch magnitude (cm)** |
| **Nerve stimulation** | |
| | |
| **Muscle stimulation** | |

*Table 8-3. Twitch duration and time to peak shortening in the frog gastrocnemius.*

| | |
|---|---|
| **Stimulus intensity:** | |
| A. Horizontal distance from onset of contraction to the peak of shortening: | |
| B. Time to peak shortening (calculated from A): | |
| C. Horizontal distance from beginning to end of twitch: | |
| D. Twitch duration (calculated from C): | |
| E. Predicted minimum stimulus frequency to produce summation (calculated from D): | |
| F. Predicted minimum stimulus frequency to produce tetanus (calculated from B): | |
| G. Results of stimulating the frog gastrocnemius at: | V for 2 sec at various frequencies. |

| Stimulus frequency (stimuli/sec) | Result |
|---|---|
| 2 | |
| 5 | |
| 20 | |
| 60 | |
| 200 | |

| | |
|---|---|
| H. Actual minimum stimulus frequency to produce summation (determined from G): | |
| I. Actual minimum stimulus frequency to produce tetanus (determined from G): | |

*Table 8-4. Response of the frog gastrocnemius to repetitive stimulation.*

| Stimulus intensity: | Stimulus frequency: |
| --- | --- |
| **Treatment** | **Result** |
| Continuous stimulation for 1 min: | |
| 5 min rest followed by continuous stimulation for 0.5 min | |

# Questions

1. What is the neurotransmitter released at neuromuscular junctions?

   _____

2. Since the action potentials of individual nerve and muscle cells are all-or-none events, what was responsible for the increase in shortening that occurred as you increased the stimulus intensity from the threshold to the optimum value?

3. Was the stimulus required to produce a response in the gastrocnemius muscle when the sciatic nerve was stimulated greater than, less than, or the same as the stimulus required to activate the muscle when the muscle was stimulated directly?

   _____

4. Explain the difference in the threshold measured for nerve stimulation and that measured for direct stimulation of the muscle.

5. Why is the maximum degree of shortening attained during a tetanus greater than that seen during a twitch?

6. a) At what frequency did summation start to occur?

   _____

   b) At what frequency did tetanus occur?

   _____

c) How did your findings, described in your answers to a and b, compare with your predictions?

7. a) Was the muscle able to sustain continuous contraction for one minute? _____ Explain.

b) Did the preparation recover after it fatigued?

8. Compare the pattern of impulses delivered to the muscle cells by the stimulator in this experiment to the pattern of activity in motor nerves when a skeletal muscle is activated by the animal's central nervous system. In other words, in what respects is this an unnatural experiment?

9. In principle, continuous stimulation could cause fatigue by a) decreasing the nerve's ability to conduct action potentials, b) depleting the neuromuscular junctions' supply of acetylcholine, c) depleting the glycogen reserves of the muscle cells, or d) reducing the effectiveness of the contractile apparatus in some way. Can you think of experiments to test some of these possibilities?

# Contractile Mechanics of Skeletal Muscle

*Reading assignment: text 292-300*

## Objectives

 *Experimental*

1. To demonstrate the effect of muscle elasticity on twitch contractions.

2. To determine the relationship between load, shortening, and muscle work.

 *Conceptual*

After completing this exercise and the reading assignment, you should be able to:

1. Define the terms **afterloaded muscle, latent period, isometric, rest length, elasticity, series elastic component, passive tension, active tension,** and **power output.**

2. Describe how load and muscle elasticity affect muscle work.

3. Distinguish between preloaded and afterloaded contractions and between isometric and isotonic contractions.

4. Explain how a muscle's ability to do work is affected by being stretched.

5. List two factors that reduce the amount of work that can be done in an afterloaded twitch by a heavily loaded muscle.

## Background

In Exercise 8 you stimulated a preloaded muscle to perform isotonic contractions. In this exercise, you will work with an afterloaded muscle performing isometric contractions. You will stimulate an isolated skeletal muscle to shorten, moving a load, and measure the effect of load on shortening.

Recall from the last exercise that a resting muscle with a load suspended from it is preloaded. When a muscle supports a load in such a way that the muscle is not under tension (when it is resting but must lift the load in order to shorten), the muscle performs an **afterloaded** contraction. In after-

loaded contractions, the tension in the muscle must rise to equal the value of the load before the load can be lifted. In other words, before the muscle can apply any energy to do work on the load, it must do some work on its own elastic elements.

The time needed for this work is largely responsible for the time between the delivery of the stimulus and the onset of contraction (**latent period**). The heavier the load, the longer the latent period (Fig. 9-1).

After a muscle lifts a load, the tension in the muscle does not change, and the contraction is isotonic. If the load is too heavy to lift, or if the muscle is

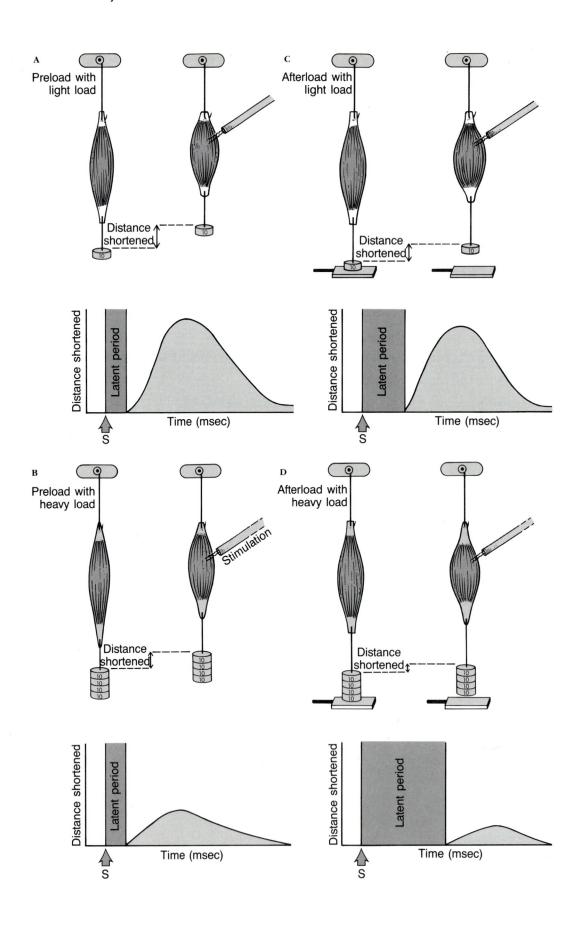

*Figure 9-1. The effect of preloading and afterloading on latent period. In a preloaded muscle, an increase in load has little effect on latent period (compare B with A). The latent period of an afterloaded muscle is longer than that of a preloaded muscle (compare C with A and D with B). Increasing the load on an afterloaded muscle causes a much larger increase in the latent period than is seen in the preloaded muscle (compare C and D with A and B). Arrows labeled S indicate application of stimuli.*

fastened between rigid supports, the muscle cannot shorten during contraction, and the contractions are said to be **isometric**. The relationship between the length of a resting muscle and its tension can be measured by attaching the muscle to two rigid supports. At one end a **force transducer** measures the tension transmitted to the support from the muscle (Fig. 9-2). Compare this method to the technique used to measure the distance a muscle shortens during an isotonic contraction (see Fig. 8-1).

The length a muscle assumes when it is detached from its connections to the skeleton is its **rest length**. **Elasticity** is a physical quality that causes a stretched muscle to resist stretching and to tend to return to its rest length when the stretching force is removed. The contractile elements of a muscle are in series with elastic components. The **series elastic components** of a resting muscle are comprised primarily of the connective tissue that surrounds muscle fibers and makes up tendons. When a load is added to the resting muscle, the elastic components of the muscle stretch until the **passive tension** (resistance to further stretching) of the muscle is equal to the force applied by the load. Passive tension is zero when a muscle is at its rest length.

In isometric contractions, the total tension present during activation is the sum of passive tension (if the muscle was stretched as it was mounted) and an additional component of **active tension** generated by the contractile machinery.

The duration of the active state of the contractile machinery is fixed for a twitch. Therefore, lengthening the latent time reduces the time available for lifting the load. In addition, the heavier the load, the less rapidly the muscle shortens after it has begun to lift the load (Fig. 9-3).

These two factors, increased latent time and reduced shortening velocity, reduce the amount of work that can be done in an afterloaded twitch by a heavily loaded muscle. In contrast to this situation, for a preloaded muscle only the shortening velocity affects the amount of work that can be done, since series elastic elements are already stretched when the twitch begins.

The amount of work done by a contracting muscle is defined as:

$$\frac{\text{mass of load (g) X distance}}{\text{load is moved (cm)}} = \text{work done (g-cm)}.$$

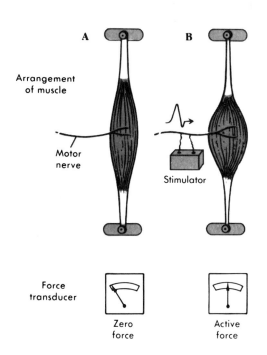

*Figure 9-2. Stimulating a muscle and recording the resulting force during an isometric contraction. A An unstimulated muscle is fixed between two supports at its resting length. B The muscle is stimulated, producing an active force that can be measured by a transducer.*

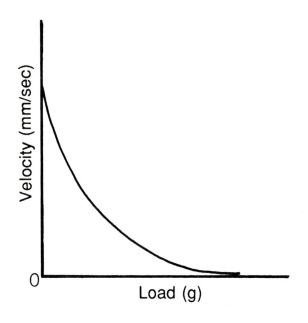

*Figure 9-3. Relationship between shortening velocity and load for a twitch.*

A muscle's rate of doing work is its **power output**. The same power output could be used to move a small load rapidly or to move a large load slowly. You will see that, for any given muscle, there is an optimal load at which the work performed in a single twitch is maximal. When muscles are activated repeatedly over time, maximum power output is obtained when the load is optimal.

## Procedure

Record your results in the laboratory report section at the end of this exercise.

 **Part 1. Setting up the muscle preparation**

1. You will use the same apparatus for mounting and stimulating the muscle and recording contractions that you used in Exercise 8. Be sure you know the leverage factor for your apparatus. The leverage factor for the muscle module is the factor by which the pen excursion must be multiplied to determine the amount of muscle shortening from the chart record. If this information is not provided, you can determine this conversion factor as follows:

Mount a metric ruler next to the muscle lever. Beginning at the point where the muscle will be attached, lift the lever several tenths of a cm and note the corresponding rise of the pen.

Your conversion factor is:

$$\frac{\text{distance moved by lever (cm)}}{\text{distance moved by pen(cm)}}.$$

Thus, the distance the muscle shortens, that is, the distance the load is moved, (in cm) is approximately equal to:

$$\text{distance moved by pen (cm)} \quad X \quad \text{conversion factor}.$$

This method is only an approximation, because it ignores the fact that the pen moves in an arc rather than in a straight line. However, for the distances you will be working with, the error introduced by this approximation will be small.

Enter the conversion factor in Table 9-1.

2. Make a frog gastrocnemius muscle preparation as in Exercise 8, except that the sciatic nerve need not be preserved, since the muscle will now be stimulated.

3. The muscle will be stimulated by two fine wires. Wrap one around the head of the muscle so that it makes good contact.

4. Fasten the second wire by a fine clip lead to the wire that attaches the Achilles tendon to the muscle lever.

5. Make sure the wires do not interfere with the movement of the muscle or the muscle lever.

6. Attach a weight pan to the muscle lever directly under the muscle.

 *Part 2. Afterloaded contractions*

1. Locate the afterloading screw (see Fig. 8-2). This can be tightened to support the muscle lever in a horizontal position. Adjust the screw so that the muscle is at its rest length and the lever is horizontal. The muscle can now lift the lever when it shortens, but the load cannot stretch the muscle. Adjust the afterloading screw and the muscle clamp so that the resting muscle is neither stretched nor slack.

2. Set the stimulator to deliver single stimuli.

3. Find the optimal stimulus intensity and record this value in Table 9-1. Leave the stimulator at this intensity for the rest of Part 2.

4. Turn off the paper advance.

5. If the weight of the pan is not known, weigh it and record the pan weight in Table 9- 1.

6. With the weight pan empty, stimulate the muscle and record the twitch that results. Because the paper is not moving, you will be recording only in one dimension (the vertical axis of the chart recording). Enter this value in

Table 9-1A. This will give you information about twitch magnitude.

7. Label each twitch with the total weight lifted. The total weight lifted is the sum of the weights added to the pan (zero in this instance) plus the pan weight.

8. Add a 10-g weight to the pan, advance the paper, and record another twitch. Label the recording and enter the height of the twitch in Table 9-1A.

9. Continue adding 10-g weights, recording twitches, and entering data in Table 9-1A until you arrive at a weight that the muscle cannot lift.

10. Set the stimulator for multiple stimuli at a frequency that should result in a tetanus (at least 60/sec). Stimulate the muscle for 2 sec. Record your results in Table 9-1B.

11. Remove the weights from the pan. Rinse the muscle with Ringer's solution and let it rest for at least 5 min.

 *Part 3. Preloaded contractions*

1. Disengage the afterloading screw, so that the muscle supports the weight of the pan.

2. Reset the stimulator to deliver single stimuli.

3. Repeat the procedure used with preloaded twitches, but add 20 g of weight at each step rather than 10 g. Record your results in Table 9-2A. After each addition of weight, return to the baseline by raising the bone clamp (refer to Fig. 8-2).

4. When you arrive at a load that the muscle cannot lift in response to a single stimulus, reset the stimulator for multiple stimuli and stimulate for 2 sec, as you did with the afterloaded muscle in Part 2, Step 9. Record your results in Table 9-2B.

# *Laboratory Report*

## *Exercise 9: Contractile Mechanics of Skeletal Muscle*

Name:

Date:

Lab Section:

## *Analyzing Your Data*

The distance the load was moved during each twitch is calculated by multiplying the height of each twitch record by the conversion factor. Plot work done as a function of load for an afterloaded and a preloaded muscle.

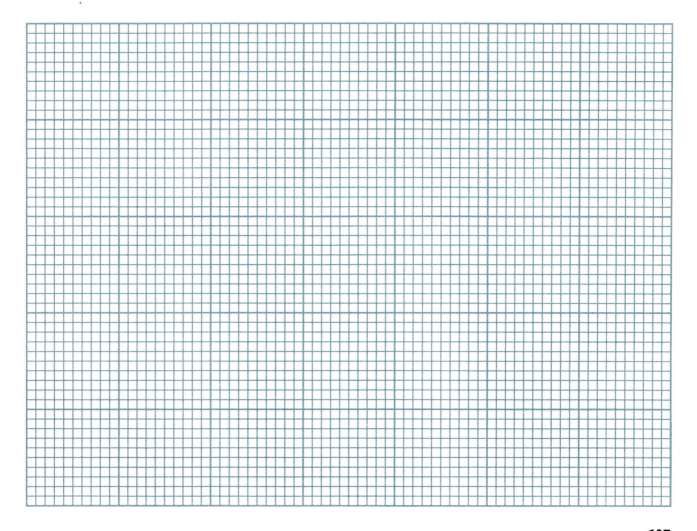

*Table 9-1. Work performed by an afterloaded gastrocnemius muscle.*

## A. Relationship between weight lifted and work performed.

Conversion factor:                         Stimulus intensity:

| Pan wt. | Load added wt.(g) | Total wt.(g) | Recorded twitch height (cm) | Distance load was moved (cm)[1] | Work performed (g-cm) |
|---|---|---|---|---|---|
| | 0 | | | | |
| | 10 | | | | |
| | 20 | | | | |
| | 30 | | | | |
| | 40 | | | | |

[1] Amount of muscle shortening (cm) = distance moved by pen (cm) X conversion factor.
[2] Mass of load (g) X distance load is moved (cm) = work done (gcm).

## B. Response of afterloaded muscle to multiple stimuli.

Stimulus frequency:                         Response of muscle:

*Table 9-2. Work performed by a preloaded gastrocnemius muscle.*

## A. Relationship between weight lifted and work performed.

Conversion factor:                     Stimulus intensity:

| Pan wt. | Load added wt.(g) | Total wt.(g) | Recorded twitch height (cm) | Distance load was moved (cm)[1] | Work performed (g-cm) |
|---------|-------------------|--------------|------------------------------|----------------------------------|------------------------|
|         | 0                 |              |                              |                                  |                        |
|         | 10                |              |                              |                                  |                        |
|         | 20                |              |                              |                                  |                        |
|         | 30                |              |                              |                                  |                        |
|         | 40                |              |                              |                                  |                        |

[1] Amount of muscle shortening (cm) = distance moved by pen (cm) X conversion factor.
[2] Mass of load (g) X distance load is moved (cm) = work done (gcm).

## B. Response of afterloaded muscle to multiple stimuli.

Stimulus frequency:                     Response of muscle:

## *Questions*

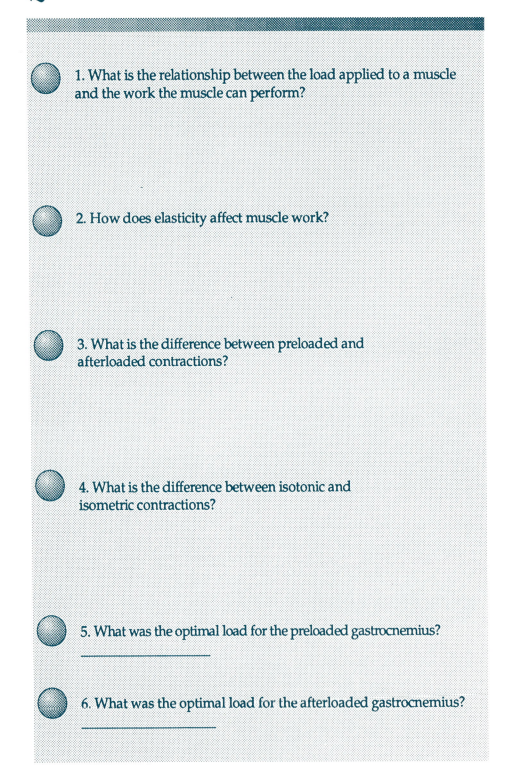

1. What is the relationship between the load applied to a muscle and the work the muscle can perform?

2. How does elasticity affect muscle work?

3. What is the difference between preloaded and afterloaded contractions?

4. What is the difference between isotonic and isometric contractions?

5. What was the optimal load for the preloaded gastrocnemius?

_____

6. What was the optimal load for the afterloaded gastrocnemius?

_____

7. How does a preloaded muscle during rest differ from a resting afterloaded muscle?

8. What factors could have caused the twitch height of the afterloaded muscle to decrease as the load was increased?

9. Could the decrease in twitch height that occurred in the afterloaded muscle as load was increased have been explained entirely by fatigue? _____ Explain.

10. What results rule out the possibility that the ultimate failure of the afterloaded muscle to lift the load as weights were added was due to an inability to generate enough active tension to lift the load?

11. The preloaded muscle was stretched as weights were added to its load. What factors could have caused the twitch height of the preloaded muscle to fall as its load was increased?

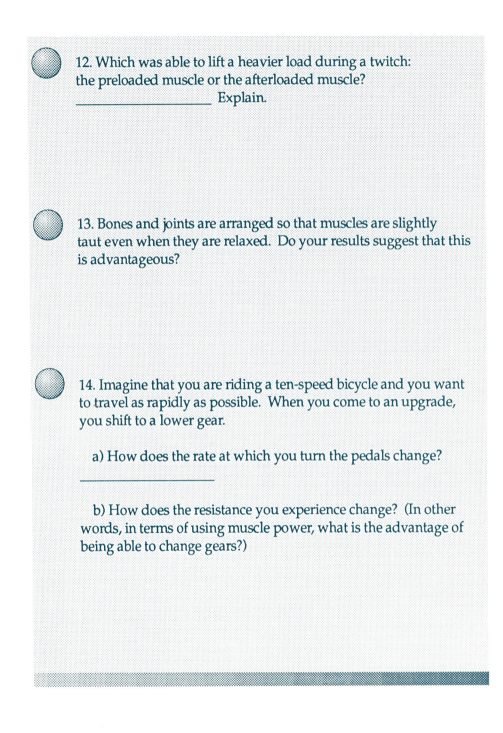

12. Which was able to lift a heavier load during a twitch: the preloaded muscle or the afterloaded muscle? _____ Explain.

13. Bones and joints are arranged so that muscles are slightly taut even when they are relaxed. Do your results suggest that this is advantageous?

14. Imagine that you are riding a ten-speed bicycle and you want to travel as rapidly as possible. When you come to an upgrade, you shift to a lower gear.

   a) How does the rate at which you turn the pedals change?

   _____

   b) How does the resistance you experience change? (In other words, in terms of using muscle power, what is the advantage of being able to change gears?)

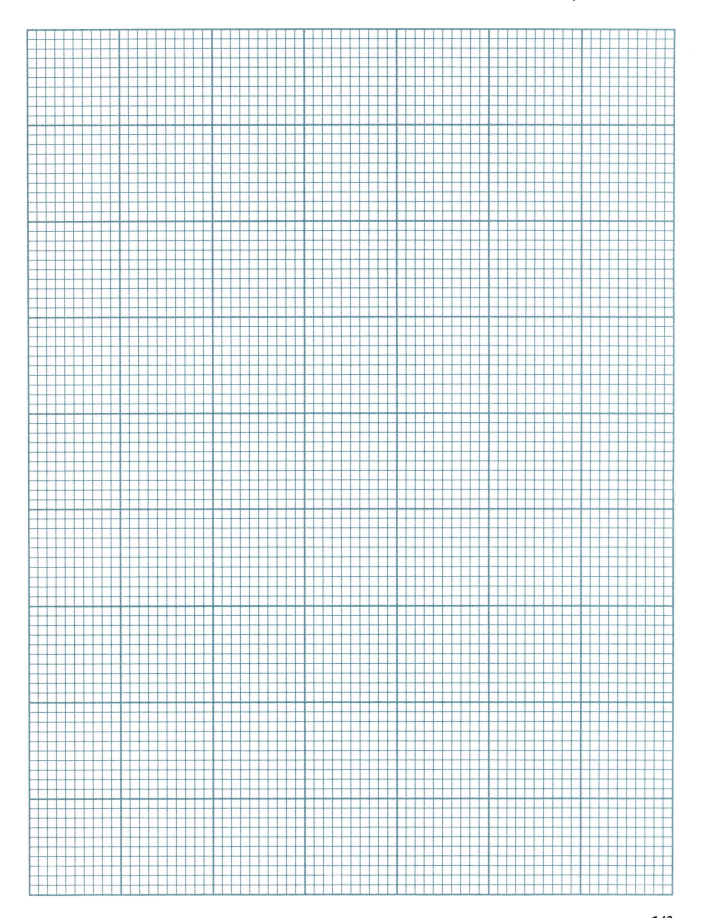

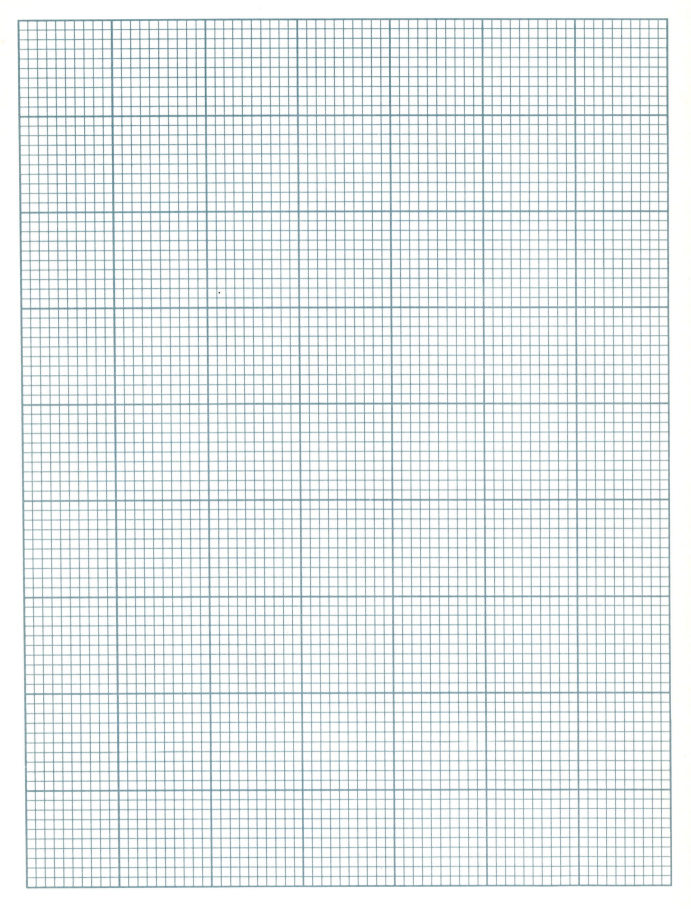

# Properties and Control of Intestinal Smooth Muscle

*Reading assignment: text 73, 259-260, 277-279, 309-314, 542-547*

## Objectives

### Experimental

1. To observe the spontaneous contraction of intestinal smooth muscle in the absence of nervous or hormonal input.

2. To determine the effects of a) epinephrine, b) acetylcholine, and c) atropine on intestinal smooth muscle.

3. To determine whether contractions of intestinal smooth muscle depend on $Ca^{+2}$ in the external solution.

### Conceptual

After completing this exercise and the reading assignment, you should be able to:

1. Compare the structures of skeletal and smooth muscle.

2. Define the terms **muscle fiber, myofibril, sarcomere, striated, pacemaker activity, slow wave, gap junction, antagonist,** and **jejunum.**

3. Compare the characteristics of single-unit smooth muscle and multiunit smooth muscle and give three examples of each.

4. Explain why single-unit smooth muscle contracts spontaneously in the absence of nervous or hormonal inputs.

5. List the two branches of the autonomic nervous system and state the neurotransmitters produced by each.

6. Describe the effects of sympathetic and parasympathetic inputs on intestinal motility.

7. Describe the **muscle bath** apparatus used to study contractions in smooth muscle.

8 Describe the effect of a) epinephrine, b) acetylcholine, and c) atropine on rabbit small intestine and explain the reasons for these effects.

## Background

In Exercises 8 and 9 you worked with **skeletal muscle**. This exercise contains a series of experiments on rabbit small intestine designed to introduce you to some of the special properties of **smooth muscle**. You will see that smooth muscle is very different in structure and function from skeletal muscle.

Skeletal muscle cells (**fibers**) contain long, parallel **myofibrils** made up of repeating subunits called **sarcomeres** (Fig. 10-1A). Adjacent sarcomeres are stacked "in register," creating patterns of repeating bands and lines. These are the characteristic striations seen in skeletal muscle viewed with a light microscope. Because of these striations, skeletal muscle is said to be **striated**. A sarcomere is a cylindrical structure containing **thick** and **thin**

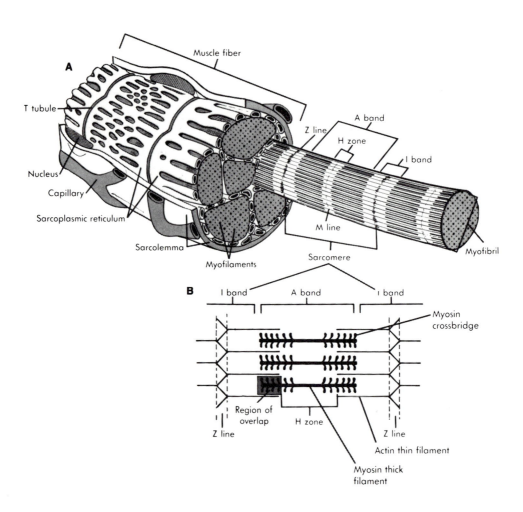

*Figure 10-1.* **A** *Organization of a striated muscle fiber. The contractile elements of a muscle fiber are the myofibrils, each composed of many repeating subunits called sarcomeres.* **B** *Within a sarcomere, thousands of individual actin and myosin molecules form a regular array of thick and thin filaments.*

filaments, which contain the contractile proteins **myosin** and **actin**, respectively (Fig. 10-1B). Skeletal muscle contraction is produced by the sliding of actin and myosin past one another.

The name *smooth muscle* is used because there are no banding patterns visible when this type of muscle is examined under the microscope (figs. 10-2 and 10-3). This is because actin and myosin are not neatly arranged in repeating sarcomeres to form myofibrils in smooth muscle. Instead, in smooth muscle, the actin and myosin filaments are oriented more or less parallel to the long axis of the cell, but they are otherwise just scattered in the cytoplasm (figs. 10-2 and 10-3).

Smooth muscle contraction is thought to result from the sliding of filaments past one another, as in striated muscle. However, much greater changes in length are possible in the loose filament arrangement found in smooth muscle.

In smooth muscle, receptors for neurotransmitters are scattered over the surface of the cell membrane. This contrasts with skeletal muscle, where receptors are clustered at the neuromuscular junction. Furthermore, unlike skeletal muscle, smooth muscle cells do not store $Ca^{+2}$ in an intricate sarcoplasmic reticulum. Instead, $Ca^{+2}$ is stored in small vesicles near the cell surface. Because they are small and contract relatively slowly compared

A  Relaxed

*Figure 10-2.* *Diagram of contraction in a smooth muscle cell.* **A** *Relaxed.* **B** *Fully contracted.* **Inset** *The relationship between thick filaments, thin filaments, and dense bodies in a single active contractile unit.*

B  Fully
contracted

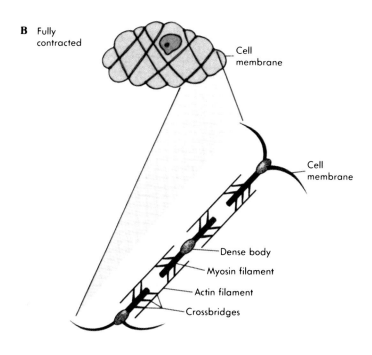

Cell
membrane

Cell
membrane

Dense body

Myosin filament

Actin filament

Crossbridges

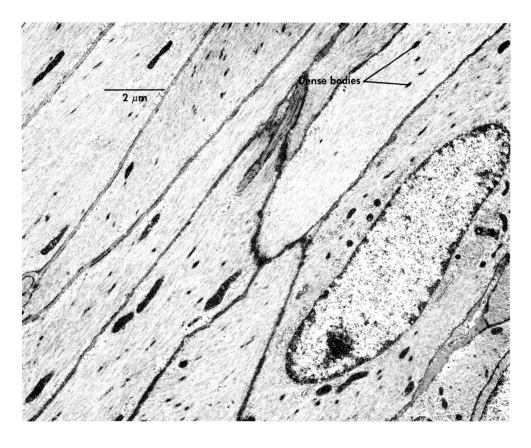

2 μm

Dense bodies

*Figure 10-3.* *Micrograph of longitudinally sectioned smooth muscle fibers. Note the absence of striations.*

to skeletal muscle cells, smooth muscle cells can obtain the $Ca^{+2}$ they need for contraction from extracellular fluid. You will have a chance to test this statement experimentally in this exercise.

Smooth muscle can be divided into two categories.

**\*\* Single-unit smooth muscle can contract in the absence of nervous or hormonal input. This type of smooth muscle is found in the walls of small blood vessels, the uterus, and the small intestine.**

**\*\* Multiunit smooth muscle does not contract spontaneously. The muscles that cause the iris to dilate and contract, that elevate body hairs (forming goose bumps), and that form sphincters in the digestive and urinary tracts are examples of multiunit smooth muscle.**

A piece of single-unit smooth muscle that is removed from the body and from all nervous connections will continue to contract. These rhythmic contractions are initiated by spontaneous, regular membrane depolarizations called **pacemaker activity** (Fig. 10-4). In most parts of the gastrointestinal tract, pacemaker activity takes the form of **slow waves** of depolarization, followed by repolarization (Fig. 10-4A). These depolarizations by themselves do not cause appreciable contraction. However, under certain conditions, a slow wave may exceed the threshold for generation of action potentials, and one or more action potentials can occur during the slow wave. These actions potentials initiate smooth muscle contraction (Fig. 10-4B). Furthermore, the activity caused by the intrinsic pacemaker can be modified by inputs from hormones, drugs, metabolic products, and the **autonomic nervous system**, the branch of the peripheral nervous system that innervates glands and smooth and cardiac muscle.

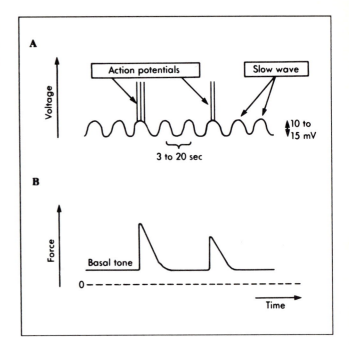

*Figure 10-4. A  Smooth muscle cells exhibit pacemaker activity called slow waves that arise in the longitudinal muscle layer. The membrane voltage varies by 10 to 15 mV, and the slow waves last 3 to 21 seconds. Bursts of action potentials generated at the top of the slow wave can initiate smooth muscle contraction. The force depends on the number of action potentials.  B  Most gastrointestinal muscle exhibits resting tension, or basal tone.*

The term "single-unit smooth muscle" refers to the fact that the activity of adjacent cells is synchronized in this type of muscle. Electrical activity is conducted from cell to cell by **gap junctions**, transmembrane channels linking the cytoplasm of adjacent cells (Fig. 10-5).

In contrast to single-unit smooth muscle, multiunit smooth muscle has few gap junctions and typically contracts when it is stimulated by synaptic transmitter chemicals or by hormones present in the blood. Because of the lack of gap junctions, neighboring cells are not synchronized. In this respect, multiunit smooth muscle resembles skeletal muscle.

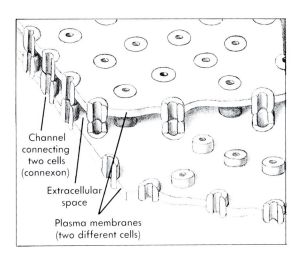

*Figure 10-5. A schematic illustration of a gap junction.*

Channel
connecting
two cells
(connexon)

Extracellular
space

Plasma membranes
(two different cells)

The rhythmic contractions of intestinal smooth muscle mix the intestinal contents during digestion and move contents from one part of the tube to another. The autonomic nervous system has dual inputs on the organs it innervates; it is subdivided into the **sympathetic** and the **parasympathetic** nervous systems, which have opposite effects. For instance, in the gastrointestinal tract, parasympathetic input increases intestinal activity. This effect is produced by the neurotransmitter **acetylcholine**. On the other hand, sympathetic stimulation, brought about by **norepinephrine**, results in a decrease in intestinal motility. The hormone **epinephrine**, produced by the **adrenal medulla** and transported to the small intestine by the circulatory system, is similar to norepinephrine in its effects. Parasympathetic inputs are inhibited by **antagonists**, drugs that interfere with receptors for neurotransmitters (in this case, acetylcholine). The drug **atropine** is an example of a parasympathetic antagonist. In this exercise you will investigate the effects of acetylcholine, epinephrine, and atropine on intestinal smooth muscle.

To be sure you understand the material that will be covered in this exercise, complete the hypotheses and the predictions below.

**Hypothesis 1:** Sympathetic input to intestinal smooth muscle causes

_____.

**Hypothesis 2:** Parasympathetic input to intestinal smooth muscle brings about

_____.

**Hypothesis 3:** The $Ca^{+2}$ necessary for contractions of intestinal smooth muscle comes from_____

**Prediction 1:** Addition of epinephrine to the smooth muscle preparation will cause

_____.

**Prediction 2:** Addition of acetylcholine to the smooth muscle preparation will cause

_____.

**Prediction 3:** Addition of atropine to the smooth muscle preparation will cause

_____.

**Prediction 4:** When the smooth muscle preparation is placed in a $Ca^{+2}$-free solution _____.

In this exercise, you will use segments of the middle portion (**jejunum**) of the intestine of a rabbit. The strip of intestine will be mounted in a **muscle bath**, consisting of an inner chamber, which is filled with **Tyrode's solution** (a physiological salt solution containing glucose for energy and several other cations and anions in addition to NaCl), surrounded by an outer chamber filled with warm water (Fig. 10-6). The muscle segments will be attached to a lever connected to a chart recorder, allowing you to keep a record of contractions.

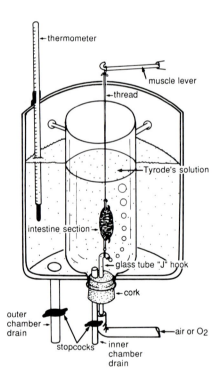

*Figure 10-6. Muscle bath apparatus for studying contractions in isolated smooth muscle.*

## Procedure

Record your results in the laboratory report section at the end of this exercise.

 *Part 1. The intestinal smooth muscle preparation*

1. Before you obtain a muscle strip, set up the muscle bath. Fill the outer chamber with water that has been warmed to $37^\circ$C to $38^\circ$C (Fig. 10-6). Fill the inner chamber with normal Tyrode's solution (that is, Tyrode's solution that contains $Ca^{+2}$) and begin bubbling air through it. Do not place the muscle in the bath until the temperature of the inner chamber has reached $37^\circ$C to $38^\circ$C.

2. Tie a short thread through one end of the intestine and fasten this to the mounting rod (glass J-hook) of the muscle holder.

3. Tie a longer thread through the mount lid and fasten this to the lever.

4. As soon as the preparation is mounted, place it in the bath. There should be no reason to take the preparation out of the chamber until each experiment is concluded. Use the drain spouts to change solutions, wash the tissue, change bath temperatures, and so forth. Use the thermometer to monitor the temperature of the outer bath throughout your experiments. If the temperature of the muscle falls below the desired range, the muscle usually will not contract. Therefore, if the temperature of the inner chamber drops below $37^\circ$C, drain and refill the outer chamber with freshly warmed water to keep the temperature of the inner bath within the desired range. When you do this, be careful not to splash tap water into the inner chamber.

5. Examine the pen of the chart recorder. Make sure you know in which direction the pen moves when the muscle shortens. Within a few minutes after the muscle is mounted, spontaneous

contractions should be visible on the chart record. Usually they are small to begin with but increase in magnitude over the following 10 to 20 min. If you look closely you will also be able to see the muscle shortening. Wait until the magnitude of the contractions is steady before beginning to experiment. If your preparation does not begin to contract within 10 to 20 min after being mounted, consult your instructor.

 *Part 2. Recording smooth muscle activity*

Be sure to label your chart recordings thoroughly and accurately so that you know exactly what treatments produced the responses you recorded. For each treatment, describe your observations, including any changes in muscle length or rate of contraction, in Table 10-1.

1. Adjust the heart lever to give a reasonable deflection and record this activity at a slow chart recorder speed for a minute or so.

2. Add about 0.1 ml epinephrine. Allow time for a response. Observe the muscle in the bath and examine the chart record.

3. Rinse the tissue and refill the inner bath with warm Tyrode's. Wait until normal activity returns.

4. Add about 0.1 ml acetylcholine. If you do not see a response, add more acetylcholine. Watch the preparation carefully. Sometimes a continuous contraction or spasm occurs, and the lever will, of course, not move when this happens.

5. Rinse the tissue and refill with warm Tyrode's. Wait until normal activity returns.

6. Rinse with $Ca^{+2}$-free Tyrode's solution and fill the chamber with $Ca^{+2}$-free Tyrode's.

7. Record 3 min of response to this treatment.

8. Add 0.1 ml acetylcholine and record for 2 min.

9. Rinse the tissue and refill with normal Tyrode's solution.

10. Observe recovery until normal activity returns.

11. Add about 0.1 ml of atropine.

12. Add acetylcholine after about 1 min and observe for 2 min.

13. Rinse thoroughly and refill with normal Tyrode's.

14. If activity does not resume after a minute or two, add some acetylcholine until the muscle responds.

15. Remove the aerator tube and observe the muscle for 5 to 15 min.

*Name:*

*Date:*

*Lab Section:*

## *Analyzing Your Data*

Your chart recordings provide a record of your data for this exercise. Be sure each diagram is clearly labeled and contains complete information on all treatments that were applied to the muscle preparation. Before you leave, go over your recordings to be sure you understand them.

Use Table 10-1 to summarize your data on the responses of the smooth muscle preparation. Then use your knowledge of muscle physiology gained from the lectures and reading assignments to interpret your data on smooth muscle responses.

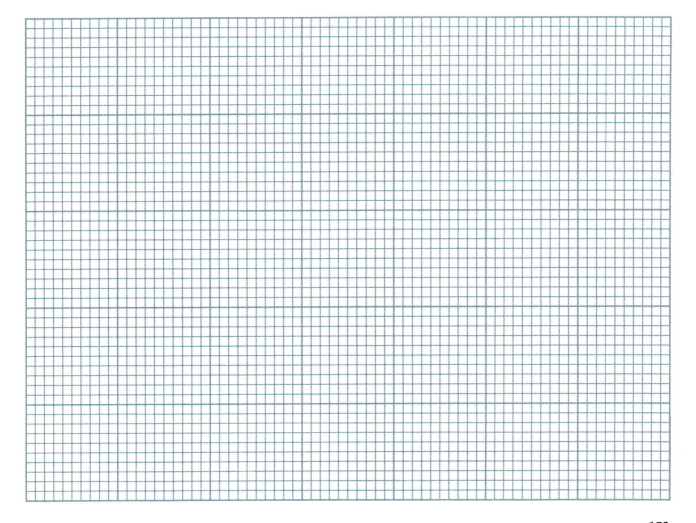

*Properties and Control of Intestinal Smooth Muscle*

*Table 10-1. Summary of rabbit small intestine activity in muscle bath.*

| Condition or treatment[1] | Responses | |
|---|---|---|
| | Contraction rate | Muscle length |
| Control | | |
| Epinephrine added | | |
| Tissue rinsed | | |
| Acetylcholine added | | |
| Tissue rinsed | | |
| $Ca^{+2}$-free Tyrode's added | | |
| Acetylcholine added | | |
| Tissue rinsed | | |
| Tyrode's with $Ca^{++}$ restored | | |
| Atropine added | | |
| Acetylcholine added | | |
| Tissue rinsed | | |
| Aerator removed | | |

[1] Preparation in 37°C to 38°C Tyrode's solution, unless otherwise indicated.

# Questions

1. Compare single-unit smooth muscle and multiunit smooth muscle.

2. Is rabbit small intestine composed of single-unit smooth muscle or multiunit smooth muscle? _____

3. What is a gap junction?

4. What role do gap junctions play in the spontaneous contractions of single-unit smooth muscle?

5. What is the effect of acetylcholine on intestinal smooth muscle? _____ Explain.

6. What effect does epinephrine have on intestinal smooth muscle? _____ Explain.

7. What effect does atropine have on intestinal smooth muscle?
_____ Explain.

8. What happened when the smooth muscle preparation was placed in a $Ca^{+2}$ free solution? _____ Why did this happen?

9. Where does the $Ca^{+2}$ that causes contraction come from in skeletal muscle? _____

10. Why is it so important to keep the smooth muscle preparation at 37°C to 38°C? Was this important when you worked with the frog muscle? Explain.

11. When you added acetylcholine to the muscle when it was in $Ca^{+2}$-free Tyrode's solution, you probably saw a brief spasm. Where did the $Ca^{+2}$ responsible for this come from?

12. What pharmaceutical applications would you expect atropine or drugs similar to atropine to have?

13. Atropine is derived from the belladonna plant. "Belladonna" means "beautiful woman." This drug was used to dilate the pupils of women's eyes, to give them a come-hither look. On the basis of this information, what can you conclude about the autonomic control of pupil diameter? _____

14. On the basis of your results, do you think contraction of intestinal smooth muscle depends on anaerobic glycolysis or oxidative metabolism? _____ Explain.

15. Considering your results, which branch of the autonomic nervous system do you suppose is most active when a meal is being digested? _____ What effect would strong emotion have on digestion? Explain.

# Sensory Physiology
# Part 1: The Visual System

*Reading assignment: text 235-250*

## Objectives

### Experimental

1. To determine the distribution of color receptors on the periphery of the retina.

2. To find out what happens when images formed on both retinas do not fall on corresponding points.

3. To demonstrate the following phenomena:
a) formation of afterimages, b) existence of a blind spot on the retina, c) accommodation for distance.

### Conceptual

After completing this exercise and the reading assignment, you should be able to:

1. Define the terms **rod, cone, central fovea, peripheral vision, corresponding points, complementary colors, optic disk, near point.**

2. Compare a) the distributions and b) the functions of rods and cones.

3. List four types of interneurons in the retina and describe their connections with each other.

4. Describe the role of mutual inhibitory connections in the perception of color.

5. Explain what causes afterimages.

6. Explain what happens when the images formed on both retinas do not fall on corresponding points.

7. Explain what causes the blind spot on the retina.

8. Describe the mechanisms by which the eye accommodates for distance.

9. Give the causes of myopia and hyperopia and explain how each of these conditions can be treated.

## Background

In this and the next two exercises, you will demonstrate several characteristics of your sensory perception and reflex responses. These exercises will give you a chance to experiment on your self, instead of using reactions in test tubes or experimental animals. Some of the phenomena you explore will be familiar; some may give surprising results.

The light sensitivity of the retina of the eye is due to a layer of photoreceptor cells called **rods** and **cones**, which are modified epithelial cells (figs. 11-1, 11-2). Rods are sensitive to wavelengths in the blue part of the spectrum. They are able to respond to low light intensity but are connected to the central nervous system so they do not give information about color. There are three families of cones: blue, green, and red cones. As their names suggest, each type is most responsive to a slightly different part of the spectrum (Fig. 11-3). The neural connections of cones allow their responses to be interpreted as sensations of color.

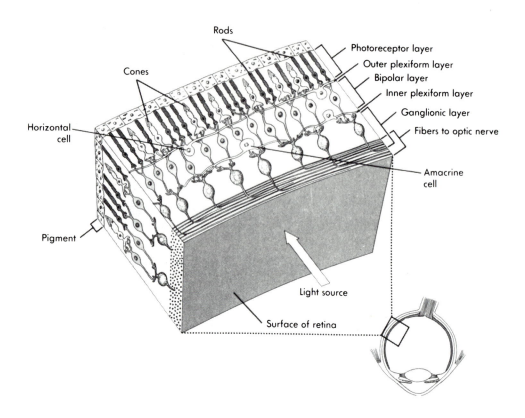

**Figure 11-1.** *A diagram of the retina. Note the bipolar cells, amacrine cells, and horizontal cells located between the photoreceptors and the ganglion cells.*

Rods and cones are not evenly distributed over the entire retinal surface. In the **central fovea,** a small part of the retina we use when we focus directly on an object to look at its color and detail (Fig. 11-4), almost all of the receptors are cones. Cones become less abundant and rods more abundant in parts of the retina that are responsible for the outer part of the visual field, or **peripheral vision.** You will demonstrate this in Part 1 of this exercise.

In addition to photoreceptors, the retina contains several types of **interneurons,** which make synaptic connections on one another. **Bipolar cells, horizontal cells,** and **amacrine cells** (Fig. 11-1) synapse on other interneurons (**ganglion cells**) that send axons from the retina to higher levels of the visual system within the brain. The axons of ganglion cells form the **optic nerve.**

The first steps of visual information processing take place in the retina. One feature of retinal processing is exaggeration of differences in light intensity and color between different parts of the image. This is the result of **mutual inhibitory connections** between retinal interneurons. Color vision is an example of this. Because of mutual inhibitory connections between interneurons in the retina, perception of a color is associated with decreased sensitivity to a second color. For example, perception of red inhibits the perception of green because red and green neural pathways in the retina make inhibitory synapses on each other. This effect enhances the contrast between the red and green parts of an image. Red and green are said to be **complementary** to one another, as are blue and yellow. When you have been fixing your gaze on a colored object for a time and then look away from it, the inhibition of the complementary

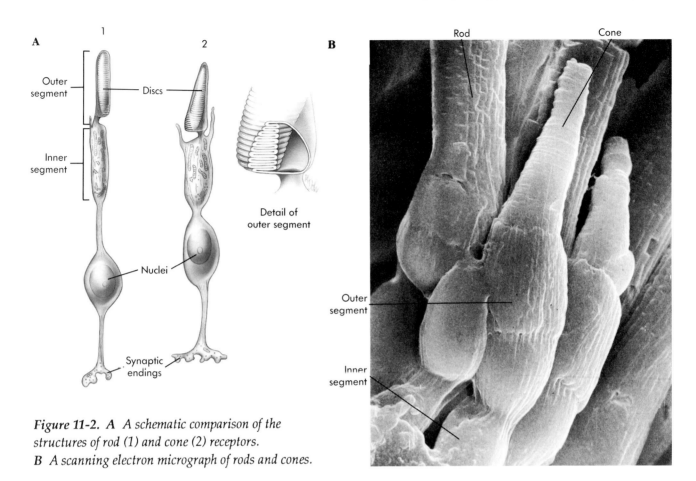

**A**

1

2

Outer
segment

Discs

Inner
segment

Detail of
outer segment

Nuclei

Synaptic
endings

**B**

Rod

Cone

Outer
segment

Inner
segment

*Figure 11-2. A  A schematic comparison of the
structures of rod (1) and cone (2) receptors.
B  A scanning electron micrograph of rods and cones.*

pathway is removed. This causes an **afterimage**, a visual hallucination which takes on the color complementary to that of the original stimulus. You will experience afterimages in Part 2 of this exercise.

When an object is viewed with both eyes open, images are formed simultaneously on the retinas of the two eyes. To avoid double vision, the two images must fall on **corresponding points** in the retinas. Receptors located at these corresponding points are connected to a single point on the brain's map of the visual field. In Part 3 of this exercise, you will discover what happens when the images fail to be formed on these corresponding points.

The axons of ganglion cells leave the retina at one point to form the optic nerve. Consequently, that portion of each retina, the **optic disk**, lacks both rods and cones and, therefore, does not form an

image. Its presence is responsible for a **blind spot** in the visual field. In Part 4 of today's exercise, you will demonstrate for yourself the existence of this blind spot.

Your eye adjusts, or **accommodates**, for light availability and also for distance. Accommodation for distance is accomplished by changes in the contour of the **crystalline lens** of the eye in response to contraction or relaxation of **ciliary muscles**.

When one looks at a distant object through the relaxed eye (corrected with glasses if need be), the lens is in a flattened state (Fig. 11-4B). As one looks at closer objects, the ciliary muscles contract and the lens becomes more rounded, or convex (Fig. 11-4A). When the lens accommodates for distance in this way, it brings the point of focus to the fovea of the retina.

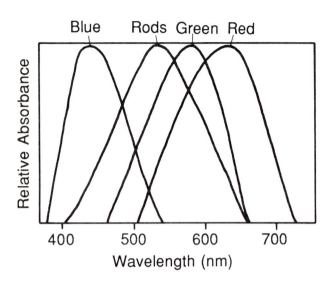

*Figure 11-3. The sensitivity of rods and three categories of cones to different wavelengths of visible light.*

The lens of the younger eye can continuously accommodate by rounding as an object comes closer and closer to the eye. If one can focus on an object brought very close to the eye, the lens is relatively elastic and is considered young. If, on the other hand, an object is not seen clearly when it is close to the eye, the person has a physiologically older lens.

The nearest point at which an object can be clearly seen is called the **near point** (Table 11-1). The distance to the near point is a measure of the relative ability of an individual to accommodate for close objects.

Although the lens and cornea of a normal eye will focus a distant object on the retina, in some individuals the eyeball is too long or too short relative to the power of the lens and cornea. If the eyeball is too long (Fig. 11-5A), the image will be focused in front of the retina; this situation is **myopia** (nearsightedness). On the other hand, if the eyeball is too short, the condition is called **hyperopia**, or farsightedness (Fig. 11-5C). Hyperopia is treated with glasses that help converge parallel rays of light (Fig. 11-5D). Myopia often occurs with a normal lens because the eyeball is too long. However, normal individuals tend to become far-

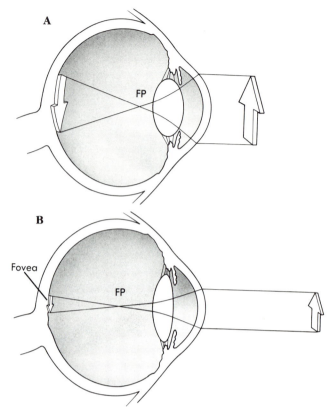

*Figure 11-4. Image formation on the retina. Note the location of the fovea, the area with the greatest photoreceptor density. FP indicates the focal point. A For near vision the ciliary muscles contract, decreasing the force exerted by the suspensory ligaments and causing the lens to become less flattened (more convex). A more convex lens bends the entering light rays more. B Distant objects are focused by a flattened (less convex) lens. The lens is flattened because the ciliary muscle is relaxed, and the suspensory ligaments are exerting their greatest force on the lens.*

sighted as they get older, since the lens loses some of its elasticity as it ages and cannot bend incoming light rays enough. Therefore, the changes of aging tend to correct myopia. Myopia is also corrected by glasses or contact lenses that cause incoming light rays to diverge slightly (Fig. 11-5B).

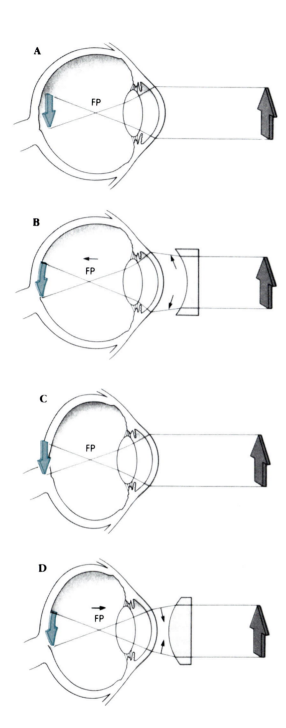

**Figure 11-5.** *Visual disorders and their correction by various lenses. FP is the focal point. A Myopia (nearsightedness) is a defect in which the cornea and lens are too powerful, or the eyeball is too long, causing the focal point to be too near the lens. The image (bluegreen arrow) is formed in front of the retina. B The effect of placing a concave lens in front of the myopic eyeball is to spread out the light rays (black arrows), moving the focal point away from the lens so that there is a focused image on the retina. C Hyperopia (farsightedness) is a disorder in which the cornea and lens are too weak or the eyeball is too short. The image is formed behind the retina. D The effect of placing a convex lens in front of the hyperopic eyeball is to bend the incoming light more, moving the focal point toward the lens.*

**Table 11-1.** *Normal values for distance to near point by age groups.*

| Age (yrs) | Distance to near point (cm) |
|---|---|
| 10 | 9 |
| 20 | 10 |
| 30 | 13 |
| 40 | 18 |
| 50 | 53 |
| 60 | 83 |
| 70 | 100 |

## Procedure

Record your results in the laboratory report section at the end of this exercise.

Work with a partner throughout this exercise. In parts 1 and 5, one of you will be the subject and the other will be the experimenter. Be sure to alternate roles, so that each person gets a chance to be both subject and experimenter for each experiment. In the other parts of this exercise, both partners should perform all the tests.

 *Part 1. Color vision*

1. Obtain the following items:

> three colored disks (red, blue, and green), protractor with a string (approximately 0.5 m in length) attached at a medial point on the straight side of the protractor.

2. Subject: hold the protractor in a horizontal position with its straight side against your forehead, just above your eyes. The protractor <u>must</u> be held in precisely the same position, and you must stare at some object directly in front of you throughout the tests. Keep one eye closed throughout the test.

3. Experimenter: attach the blue disk to the free end of the string. Pull the string laterally until it is taut. It should cross the 180° angle on the protractor. <u>Slowly</u>, move the blue disk forward and have the subject tell you

> a) when he or she first sees it and
> b) when he or she can identify its color.

4. Record (in Table 11-2) the angle at which the disk was seen and the angle at which its color was identified. Repeat the procedure using each of the disks in turn.

5. Use the blue disk once more and check the angles at which the object and its color are detected, by the same eye tested before, but with the disk held on the other side of the subject.

 *Part 2. Afterimages*

1. Place an 8-1/2 x 11″ sheet of pure white paper and another sheet of 8-1/2 x 11″ jet black paper on the laboratory desk a short distance apart. Place a bright blue 3 x 3″ card on the black paper. Stare at the blue card intently for 30 sec.

2. Shift your gaze quickly to the white paper. Record your observations in Table 11-3.

3. Repeat with a yellow card.

4. Repeat again, but place the card initially on the white paper and shift your gaze to the black paper.

 *Part 3. Corresponding points*

1. With both eyes open, focus on an object in the distance. While keeping that object in focus, gently press on the lateral portion of your right eye. This moves the eye only slightly but causes the image of the object being viewed to fall on a portion of the right retina that does not quite correspond to the area on which the image is formed on the left retina. Record your results in Table 11-4.

2. While you continue to press on the side of your right eye and view a distant object, close your left eye. Next open your left and close your right eye. Record your observations in Table 11-4.

 *Part 4. The blind spot*

1. Cover the left eye and focus the right eye on the cross in Fig. 11-6. The paper should be positioned so that the dot is to your right and the cross is directly in front of the right eye.

*Figure 11-6. Images for detection of blind spot.*

2. Beginning with the paper at arm's length, bring it slowly toward the eye until a point is reached at which the cross is seen but the dot is not visible.

3. Once more, with the left eye covered, look at the cross on the paper and position the paper so that the image of the dot falls on the optic disk and is not visible. Now, without moving the paper or your head, uncover the left eye. Record your observations in Table 11-5.

 *Part 5. Accommodation for distance*

1. Subject: If you wear glasses, keep them on for this experiment. Close one eye. Hold a pencil at arm's length, and gradually bring it closer to the opened eye. Stop when the pencil goes out of focus.

2. Experimenter: use a ruler to measure the distance from the pencil to your eye. Record this value in Table 11-6 and on the board. Compare your values to the normal values for your age group listed in Table 11-1.

# Laboratory Report

Name: _____

Date: _____

Lab Section: _____

## Analyzing Your Data

In this exercise your data analysis will consist of interpretation of data that are primarily qualitative or descriptive rather than quantitative. Review your observations and the reading assignment; then answer the questions at the end of this exercise.

Use the pooled class data from Part 5 of this exercise to plot distance to the near point, in cm, as a function of age, in years.

*Table 11-2. Peripheral vision: detection of colored disks.*

| Color | Position of disk | Angle at which disk was detected | Angle at which color was perceived |
|---|---|---|---|
| Red | Right | | |
| | Left | | |
| Blue | Right | | |
| | Left | | |
| Green | Right | | |
| | Left | | |

**Table 11-3.** *Formation of afterimages using different colors for object viewed and background.*

| Color of card | Color of background paper | Color of afterimage |
|---|---|---|
| Blue | black | |
| Yellow | black | |
| Yellow | white | |

**Table 11-4.** *Appearance of an object viewed while pressing gently on right eye.*

| Object viewed with | Appearance of object[1] |
|---|---|
| Both eyes open | |
| Right eye only | |
| Left eye only | |

[1] Single or double image.

*Table 11-5. Detection of retinal blind spot.*

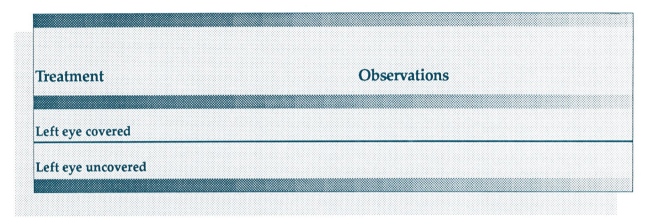

| Treatment | Observations |
|---|---|
| Left eye covered | |
| Left eye uncovered | |

*Table 11-6. Relationship between accommodation for age and distance.*

| Subject | Distance to near point (cm) | Age (yrs) |
|---|---|---|
| 1 | | |
| 2 | | |

## Questions

1. a) Which type of your cone cells do your data indicate are located most peripherally? _____

   b) Which type of cone seems most narrow in its distribution? _____ Explain.

2. Were all colors identified at different angles? _____ Explain.

3. Can you determine from your data if rods or cones are located more peripherally in the retina? _____

4. Was the angle at which you recognized a given color the same on both sides? _____ Explain.

5. How many different receptor types do you need to account for your results in the test of peripheral vision (Part 1)? _____

6. Is it a great disadvantage to have only rods in the periphery of your receptive field? _____ Explain. (Hint: What is your response when you detect a moving object in the periphery of your field of vision?)

7. What causes afterimages?

8. a) Did you see a double image in any of these exercises? _____

   b) If so, what caused the double images?

9. What causes the blind spot?

10. Why are you usually unaware of your blind spot?

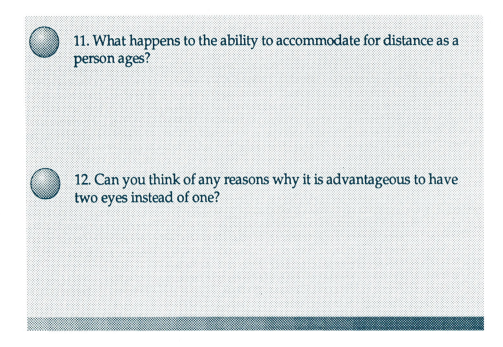

11. What happens to the ability to accommodate for distance as a person ages?

12. Can you think of any reasons why it is advantageous to have two eyes instead of one?

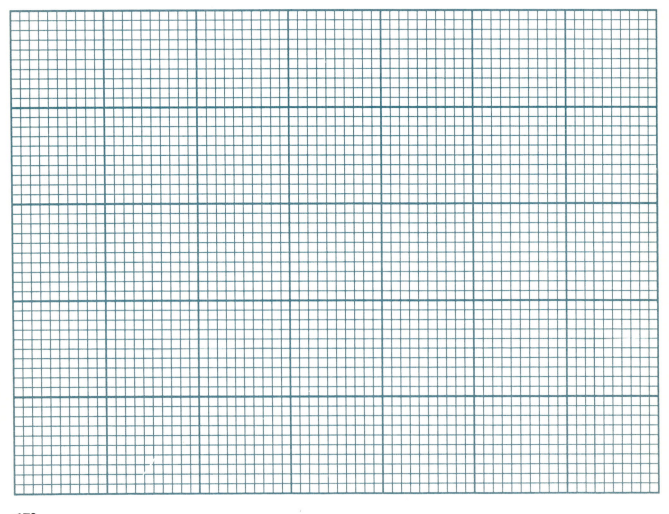

# Sensory Physiology
## Part 2: The Sense of Touch and the Chemical and Auditory Senses

*Reading assignment: text 204-212, 223-235, 250-253*

## Objectives

### Experimental

1. To compare sound perception when a vibrating tuning fork is a) held near the ear and b) in contact with the skull.

2. To test the student's ability to localize sound.

3. To determine the effect of a) visual, b) tactile, and c) vestibular inputs on the sense of balance.

4. To determine the distributions and thresholds of receptors for the four primary tastes.

5. To observe the effect of immediate prior experience on the perception of a) taste and b) temperature.

6. To determine the two-point thresholds of touch receptors in the skin of various parts of the body.

### Conceptual

After completing this exercise and the reading assignment, you should be able to:

1. Describe the mechanism by which sound waves are conducted by the middle ear.

2. List two types of sensory clues that aid in sound localization and discuss the conditions under which each is important.

3. List three types of sensory input that contribute to the ability to maintain balance.

4. Describe the mechanisms by which acceleration and rotation are detected by the inner ear.

5. Define **rotational nystagmus**.

6. List the four primary tastes.

7. Describe the relationship between variation in the density of touch receptors in the skin and representation in the sensory cortex.

## Background

Sensations fall into categories or **sensory modalities,** such as touch, taste, temperature, pain, hearing, vision. Some sensory modalities, for example vision and hearing, are mediated by specialized organs and are called **special senses**. Others, such as touch, temperature, and pain, are mediated by all parts of the body surface. These are **somatosensory** modalities. Exercise 11 investigates the visual system — the sense of sight. This exercise deals with other special senses (hearing, the sense of balance, and the chemical senses, taste and smell) as well as somatosensory modalities.

In this exercise, you will observe your own sensations and try to understand them. In Part 1 you will test your perception of the sound made by a vibrating tuning fork. Usually, perception of a sound occurs when sound waves pass through the air and strike the **tympanic membrane** (Fig. 12-1). The sound energy is transmitted to the **auditory ossicles** of the middle ear and then to the **oval window,** which transmits it to auditory receptors in the **cochlea** of the inner ear. However, sound can also be conducted directly to the cochlea by bones of the skull, as when you touch the handle of a vibrating tuning fork to your skull. If the sound is heard when the tuning fork is in contact with the skull but not when it is vibrating in the air near the ear, this may indicate a failure of the ear to conduct sound waves normally via the tympanic membrane and ossicles, or it may be due to occlusion of the auditory canal.

In Part 2 of this exercise you will test your ability to localize sound. A person with normal hearing can generally determine the direction from which a sound emanates. Two basic clues make this possible, but you must have two functioning ears to make use of either of them. One clue is the **time of arrival** of the sound waves at each ear. Earlier arrival of a sound wave at one ear indicates the sound source is on that side of the head (Fig. 12-2A). Another clue is given by the **difference in intensity** of a sound as perceived by each ear (Fig. 12-2B). Of these, the time-of-arrival clue seems to be most important for low-pitched sounds and the intensity clue the most important for sounds of high pitch.

Several types of sensory input contribute to the ability to maintain balanced erect posture, including inputs from the eyes, receptor cells in the vestibular system of the inner ear, and muscle receptor organs that are involved in maintaining

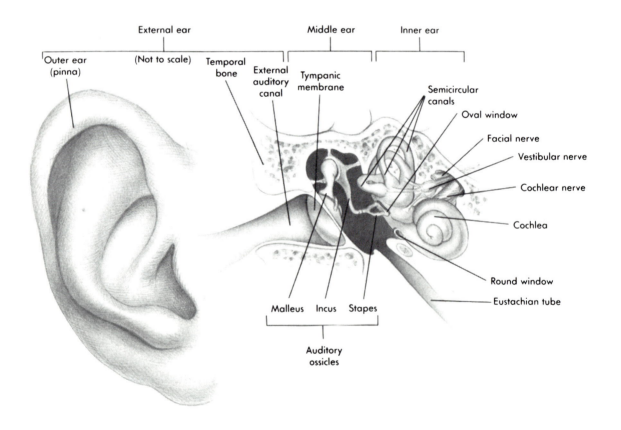

*Figure 12-1. View of the external ear, middle ear, and inner ear. The structures shown are not drawn to scale.*

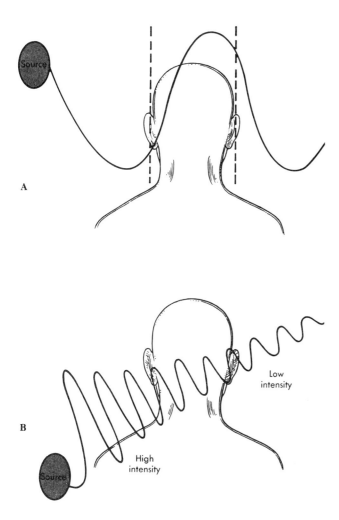

*Figure 12-2. A Sound localization clues are provided by phase differences for sounds of low frequency, in which each pressure wave strikes the right ear and the left ear at slightly different times. B For sounds of high frequency, phase differences would give ambiguous information because there is time for more than one wave cycle while the sound travels the width of the head. Thus the intensity difference must be used as a directional clue. The head absorbs some of the sound, emphasizing the intensity difference.*

posture. You will demonstrate the roles of vision and the vestibular system when you experiment with your sense of equilibrium in Part 3.

Vision gives information about the orientation and movement of the body in relation to objects in the visual environment. Another kind of environ-

mental clue comes from muscle and tendon receptors, which give information about the distribution of gravitational forces. Differences in loading and tension between the two legs, for example, occur when the body is off balance. Finally, the **vestibular system** of the inner ear gives information about the orientation of the head in relation to the pull of gravity and acceleration or deceleration of the head.

Head position and linear acceleration are sensed by the **saccule** and **utricle**, small chambers in each inner ear lined with hair cell receptors (Fig. 12-3A). Each chamber contains crystals of calcium carbonate called **otoliths** that rest on and stimulate **receptor hair cells** (Fig. 12-3B). When the head changes its orientation, the otoliths move, and a different population of receptors is stimulated.

Rotation is detected by three **semicircular canals** in each inner ear (Fig. 12-3C). Each of these fluid-filled loops ends in an **ampulla**, a swelling at the end closest to the utricle (Fig. 12-3B). The ampullae contain brush-like tufts of receptor hair cells that extend into a ball of gelatinous material, called the **cupula** (Fig. 12-3C). The canals are oriented so that they can detect rotation of the head in three planes. When the head rotates, the inertia of the fluid in the canals in the plane of rotation causes it to lag behind the movement of the canal walls, bending the cilia of the hair cells and stimulating them. The otoliths and semicircular canals normally allow balance to be maintained (Fig. 12-4).

Vestibular input can influence eye movement. This allows you to keep your gaze fixed on a point in the visual field even if your head is moving. If you sit on a stool that is rotated and then stopped, the fluid in your semicircular canals continues to rotate after the stool stops moving. During this time, your eyes rotate in the direction in which the fluid is moving, as if trying to keep up with the perceived movement of your head, then snap back rapidly and move in the original direction again. This reflexive eye movement is called **rotational nystagmus**.

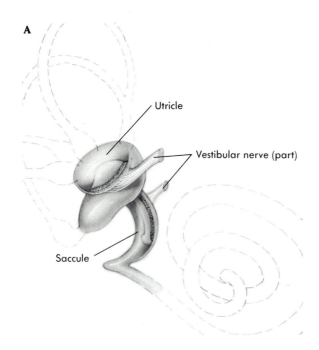

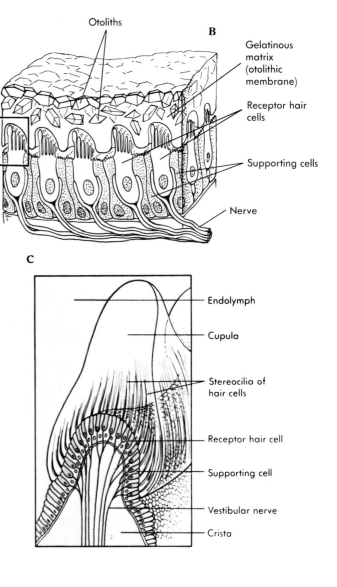

*Figure 12-3.* **A** *The position of the saccule and utricle in relation to the semicircular canals of the inner ear.* **B** *Enlargement of a section of the utricle showing otoliths and receptor hair cells.* **C** *Enlargement of a section of an ampulla showing receptor hair cells.*

Parts 4 and 5 of this exercise deal with taste, one of the chemical senses. The sense of taste originates in **taste buds**, small clusters of primary receptor cells on the tongue, the roof of the mouth, and the palate (Fig. 12-5). The receptors in taste buds are modified epithelial cells. There are four **primary flavors**: sweet, salty, sour, and bitter. Most of the subjective sensations of flavor that we get from foods are the result of smell rather than taste. Sugars and amino acids are typically sweet, acids are sour, and ionized solutes are typically salty. Many poisons are bitter, suggesting that the unpleasantness associated with bitter flavor may have adaptive significance in helping us avoid potentially harmful substances.

The term **threshold** can have several meanings in physiology. When used in the context of sensory perception, it refers to the minimum intensity of a stimulus that can be consciously perceived. The threshold for perception of a stimulus is determined by the response characteristics of the receptors and by the responsiveness of the central

*Figure 12-4.* *Function of otoliths and semicircular canals in maintaining balance.* **A** *The utricle responds to changes in the position of the head relative to gravity. When the head is upright there is a uniform downward force on the hair cells.* **B** *Tilting the head forward causes a shearing force.* **C-E** *The semicircular canals respond to rotations. As a person begins to spin, the cupula is displaced in a direction opposite to the direction of spin. This displacement exerts a shearing force on the hair cells. When the spinning stops, the inertia of the fluid causes the cupula to be displaced in the same direction as the original spin.*

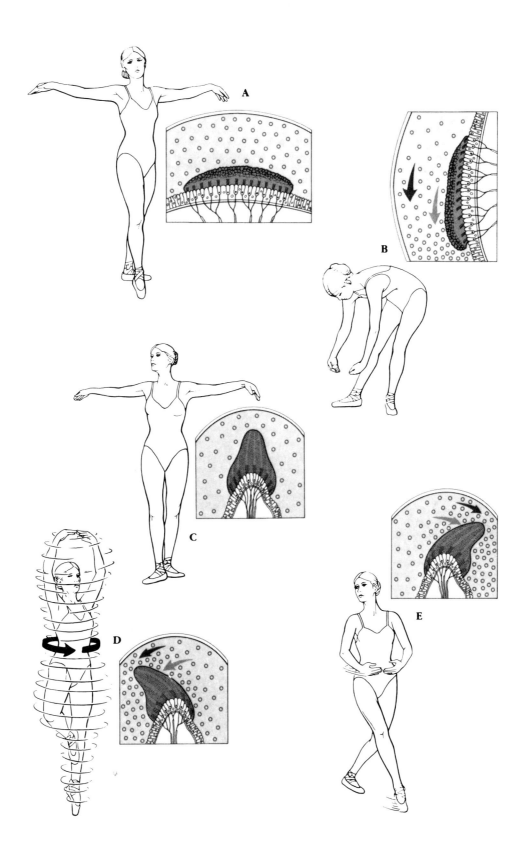

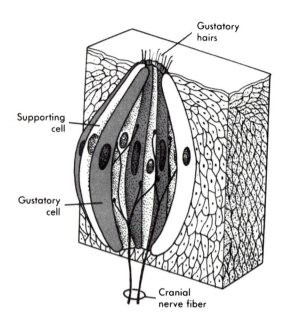

*Figure 12-5. Diagram of a taste bud.*

One measure of the somatosensory system is the **two-point discrimination test**. In this test (Part 7), you will test your ability to discriminate between two points on your skin that are stimulated simultaneously (Fig. 12-6) and compare this to a map of cortical sensitivity to pressure over different parts of the body. In this map of the receptive fields of the cerebral cortex, termed the **somatosensory homunculus**, the most sensitive parts of the body (such as the fingers and lips) have the greatest density of receptors (Fig. 12-7).

nervous system. Both of these are affected to a certain extent by the recent experience of the person being tested.

Not only can one flavor alter the perception of others, but many flavor stimulants can cause taste hallucinations if they are followed by pure water, a substance that has no taste of its own. These aftertastes are analogous to visual afterimages that are seen when a colored stimulus is followed by a white stimulus. They are examples of how the responsiveness of a sensory system is conditioned by previous experience.

The last two parts of this exercise involve the somatosensory modalities pressure (Part 5) and temperature (Part 6). The perception of temperature provides another demonstration of the influence of prior experience on sensory perception.

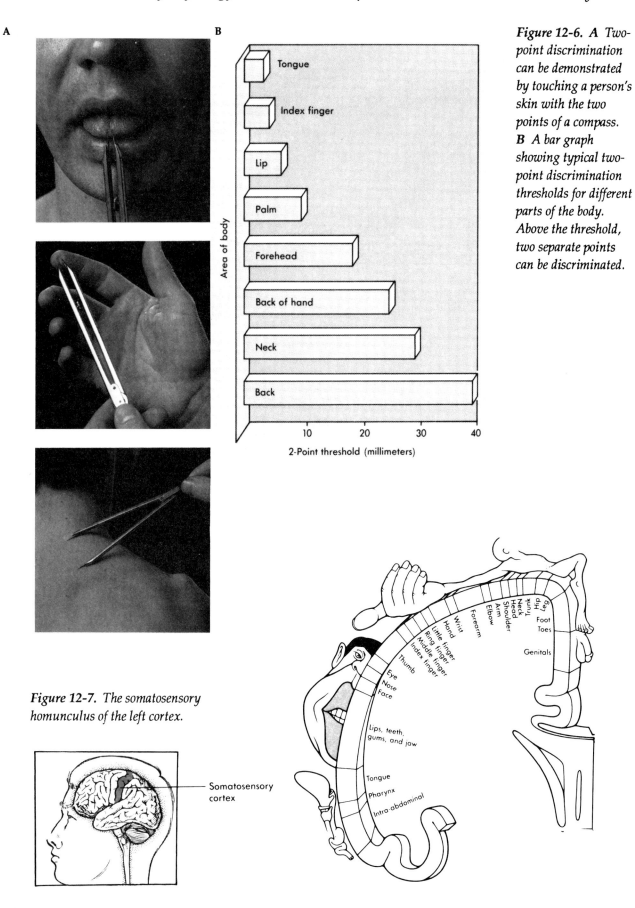

**A**

**B**

**Figure 12-6. A** *Two-point discrimination can be demonstrated by touching a person's skin with the two points of a compass.* **B** *A bar graph showing typical two-point discrimination thresholds for different parts of the body. Above the threshold, two separate points can be discriminated.*

Area of body

Tongue
Index finger
Lip
Palm
Forehead
Back of hand
Neck
Back

10    20    30    40

2-Point threshold (millimeters)

**Figure 12-7.** *The somatosensory homunculus of the left cortex.*

Somatosensory cortex

179

## Procedure

Record your results in the laboratory report section at the end of this exercise.

Work with a partner again. In parts 2, 3, 4, and 7 you will take turns being the subject and the experimenter. Both partners (Subject 1 and Subject 2) should perform all tests. In parts 1, 5, and 6 you will each do the experiment.

 ### Part 1. Transmission of sound to the inner ear

1. Strike a tuning fork against the palm of the hand to make the fork vibrate.

2. Hold the fork so that it is just outside the external ear but not in contact with your skull. Record your results in Table 12-1.

3. Touch the handle of the vibrating fork to the process of the temporal bone that projects behind the ear. Record your experience in Table 12-1.

 ### Part 2. Localization of sound

1. Subject: close your eyes.

2. Experimenter: strike the tuning fork again and hold it so that it is not in contact with the subject's head. Use the vibrating tuning fork to produce a sound in various positions (above, below, to the right, to the left, or directly in front of the subject's head) and have the subject state the direction from which the sound seems to come. Record your results in Table 12-2.

3. Repeat Step 1 with one of the subject's ears plugged with cotton. Record your results in Table 12-2.

 ### Part 3. The sense of equilibrium

1. Subject: stand on one foot with the other leg flexed at the knee. Hold your arms straight out to the side.

2. Experimenter: time how long your partner can hold this posture without moving his or her arms. Enter your results in Table 12-3A.

3. Subject: relax. Then repeat Step 1, with both eyes closed. Use Table 12-3A to record what happens.

4. Experimenter: again time how long your partner can hold this posture without moving his or her arms. Record your results in Table 12-3A.

5. Repeat steps 1 through 4 with the subject standing on one foot as before, but allow one arm to rest lightly on a table top. Enter your results in Table 12-3A.

6. Subject: sit on a swivel stool with your eyes shut and your head tilted at an angle of 30° from upright.

7. Experimenter: rotate the subject as fast as possible for 10 or 11 turns.

8. Subject: as soon as the stool has stopped spinning, open your eyes.

9. Experimenter: note any movement of the subject's eyes. Describe your observations in Table 12-3B.

10. Repeat steps 6-9. This time the subject should try to walk when the stool stops spinning. (Be careful. The subject may need support to avoid falling.) Describe your observations in Table 12-3B.

## Part 4. Distribution and thresholds of taste receptors

The test solutions will contain various concentrations of the following substances:

> NaCl (salty),
> sucrose (sweet),
> citric acid (sour),
> quinine (bitter).

The subject should be kept unaware of the composition of test solutions, and the different concentrations of test solutions should be presented in random order.

1. Subject: before each test, blot your tongue with a paper towel. When the experimenter hands you a piece of filter paper dipped into a test solution, touch it to various spots on the surface of your tongue. You may move the filter paper to different positions before concluding that a flavor can or cannot be perceived. Use Fig. 12-A to record the regions in which each taste can be most easily recognized. Rinse your mouth when you have completed the test.

2. Experimenter: obtain four strips of filter paper. Dip the first piece into the most concentrated solution of one of the test substances and give it to the subject. Record the results in Fig. 12-A.

3. Experimenter: repeat Step 2 three more times. Each time, use a fresh piece of filter paper and the most concentrated solution of a different test solution substance. Enter your results in Fig. 12-A.

4. Experimenter: repeat the taste test three more times, using the other concentrations of one of the test substances. This time, tell the subject which taste is being tested and have him or her place the strips of filter paper on the region of the tongue that is most sensitive to that flavor. Enter your results in Table 12-4.

5. Experimenter: repeat Step 4 three more times using the remaining concentrations of each of the other tastes. When you have completed this step, you should have used all concentrations of all the test substances. Enter your results in Table 12-4. (Be certain that you record your results in the proper rows.)

6. Using the results summarized in Table 12-4, determine the threshold concentration for each of the primary tastes. Circle these values in Table 12-4.

## Part 5. Aftertastes

1. Hold a mouthful of a solution of sodium bicarbonate in your mouth for two minutes. Spit out the sodium bicarbonate and take a mouthful of water. Record your experiences in Table 12-5.

2. Repeat with a solution of citric acid followed by water. Record your results in Table 12-5.

## Part 6. Sensing temperature

Both partners should do the following:

1. Obtain three 1000-ml beakers. Place about 500 ml of ice water in the first beaker, 500 ml of water at room temperature in the second beaker, and 500 ml of water at 45°C in the third beaker.

2. Place your left hand in the beaker of ice water and your right hand in the beaker of 45°C water.

3. After immersion for at least 1 min, place both hands simultaneously in the water at room temperature. Describe your sensations in Table 12-6.

## Part 7. Touch receptors: two-point discrimination

1. Subject: close your eyes while stimuli are applied or look away from the part of the body being stimulated.

2. Experimenter: obtain a pair of dividers, the tips of which can be pressed gently against the skin, to stimulate touch receptors.

3. Experimenter: gently place the divider points side-by-side on the subject's forearm. Instruct the subject to tell you when he or she can first detect two points of stimulation. Gradually move the points of the divider apart. Test the accuracy of the subject's reporting by occasionally using only a single point of stimulation.

4. When two points are discernible, measure the distance in mm between the points of stimulation and record this value in Table 12-7. This is the two-point threshold for that region of the body.

5. Repeat steps 2 to 4 two times. Enter your data in Table 12-6.

6. Calculate the average of the three values for your two-point threshold and enter this value in Table 12-6.

7. Determine the two-point thresholds for the each of the following parts of the body:

> lower part of the leg,
> sole of the foot,
> forefinger,
> lower lip.

8. Repeat each test twice. Calculate the average value for each part of the body, and enter these numbers in Table 12-7.

9. Compare your results to Fig. 12-7 showing the portions of the sensory cortex that receive stimuli from different regions of the body.

# Laboratory Report

### Exercise 12: Sensory Physiology - Part 2: The Sense of Touch and the Chemical and Auditory Senses

*Name:* _____

*Date:* _____

*Lab Section:* _____

## Analyzing Your Data

Analyze your data for this lab by reviewing your observations (use your notes, as well as the tables and Fig. 12-A) and the reading assignment before answering the questions below.

*Table 12-1. The perception of sound.*

| | Perception of sound[1] | |
|---|---|---|
| **Experimental conditions** | **Subject 1** | **Subject 2** |
| Tuning fork in contact with skull | | |
| Tuning fork not in contact with skull | | |

[1] Yes or no.

*Table 12-2. Localization of sound.*

| | Perception of sound[1] | |
|---|---|---|
| **Experimental conditions** | **Subject 1** | **Subject 2** |
| **Without ears plugged** | | |
| Tuning fork above head | | |
| Tuning fork below head | | |
| Tuning fork left of head | | |
| Tuning fork right of head | | |
| Tuning fork in front of head | | |
| **With one ear plugged** | | |
| Tuning fork above head | | |
| Tuning fork left of head | | |
| Tuning fork right of head | | |
| Tuning fork in front of head | | |

[1] Yes or no.

*Table 12-3. The sense of equilibrium.*

## A. The effect of visual and tactile inputs on balance.

| | Time until subject standing on 1 foot moved arms (sec) | |
|---|---|---|
| Experimental conditions | Subject 1 | Subject 2 |
| Eyes open | | |
| Eyes closed | | |
| Eyes open, touching table top | | |
| Eyes closed, touching table top | | |

## B. The effect of rotation on eye movements and balance.

| Responses of subject | Subject 1 | Subject 2 |
|---|---|---|
| Eye movements | | |
| Ability to walk | | |

*Figure 12-A.* *Distribution of taste receptors. Indicate which regions of the tongue perceived the test solutions by using the following symbols: S = salty, SW = sweet; SO = sour, and B = bitter.*

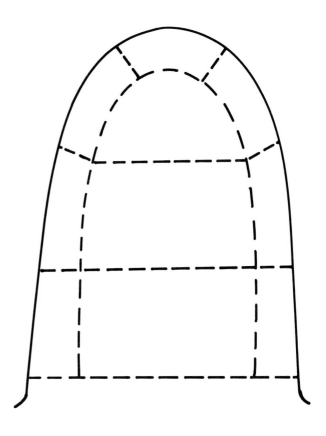

*Table 12-4. Thresholds of primary tastes.*

| | Perception of taste[1] | |
|---|---|---|
| Test concentration (M) | Subject 1 | Subject 2 |
| **Salty** | | |
| 0.0010 | | |
| 0.0032 | | |
| 0.0100 | | |
| 0.0320 | | |
| 0.1000 | | |

**Table 12-4,** *continued*

| | |
|---|---|
| **Sweet** | |
| 0.0010 | |
| 0.0032 | |
| 0.0100 | |
| 0.0320 | |
| 0.1000 | |
| **Sour** | |
| 0.00010 | |
| 0.00032 | |
| 0.00100 | |
| 0.00320 | |
| 0.01000 | |
| **Bitter** | |
| 0.0000010 | |
| 0.0000032 | |
| 0.0000100 | |
| 0.0000320 | |
| 0.0001000 | |

[1] Yes or no.

**Table 12-5.** *The effect of prior experience on the sense of taste.*

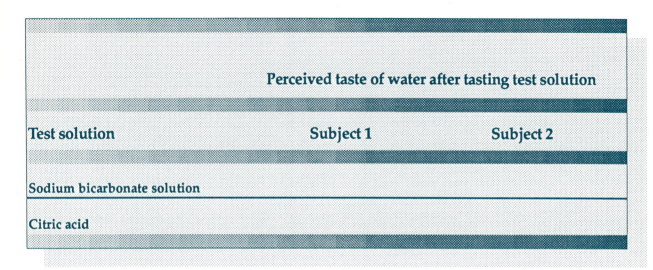

| | Perceived taste of water after tasting test solution | |
|---|---|---|
| **Test solution** | **Subject 1** | **Subject 2** |
| Sodium bicarbonate solution | | |
| Citric acid | | |

**Table 12-6.** *The effect of prior experience on the sensation of temperature.*

| | Perceived temperature[1] | |
|---|---|---|
| **Experimental conditions** | **Subject 1** | **Subject 2** |
| Left hand: ice water, followed by room temperature water | | |
| Right hand: 45°C water, followed by room temperature water | | |

[1] Warm or cool.

*Table 12-7. Two-point thresholds for different regions of the body.*

| | Two-point threshold (mm) | | | |
|---|---|---|---|---|
| **Region of body** | a[1] | b | c | average |
| forearm | | | | |
| lower leg | | | | |
| sole of foot | | | | |
| forefinger | | | | |
| lower lip | | | | |

[1] a, b, and c denote first, second, and third trials.

## Questions

1. Are two ears better than one? _____ Explain.

2. Did touching the table top make it easier to stand on one foot? _____ Why?

3. What other types of sensory input contribute to the sense of equilibrium? How?

4. How are acceleration and rotation detected?

5. What is rotational nystagmus?

6. a) Which type of receptors has the lowest threshold: receptors for sweet, salty, bitter, or sour substances? _____

   b) What might be the adaptive advantage of this difference in sensitivity?

7. How did the two-point thresholds for different parts of your body compare with those shown in Fig 12-6B and with the somatosensory homunculus?

8. State whether you agree or disagree with the following statement, giving examples to support your position: Our senses always give us a truthful representation of the physical world.

# Human Reflexes

*Reading assignment: text 86-90, 94-96, 261-268*

## Objectives

### Experimental

1. To demonstrate the following a) stretch reflexes, b) reflexes of the eye, and c) reflexes of the digestive system.

2. To observe the effects of mental activity, simultaneous muscular activity, and fatigue on a stretch reflex.

3. To determine the effects of vinegar in the mouth on the salivary reflex.

4. To determine the influence of conscious control on the swallowing reflex and the palatal (gag) reflex.

### Conceptual

After completing this exercise and the reading assignment, you should be able to:

1. Explain how a negative feedback system contributes to homeostasis.

2. List the components of a reflex arc.

3. Define the terms **reflex arc, spinal reflex, conditioned reflex, stretch reflex, myotatic reflex,** and **tendon reflex.**

4. Identify the components responsible for each of the reflexes demonstrated.

5. State the adaptive value of each of the reflexes demonstrated.

6. Discuss the role of cerebral activity in spinal reflexes.

7. Describe the process by which conditioned responses are created.

8. Define the terms **ipsilateral reflex, contralateral reflex,** and **consensual reflex** and give an example of each.

9. State the effects of parasympathetic and sympathetic inputs to the salivary glands.

## Background

A **reflex** is a sequence of events initiated by a sensory input and executed by a muscle or gland. Reflexes are the simplest functional responses of the nervous system. Reflex actions often appear purposeful. For example, the withdrawal of an appendage from a painful stimulus and the movements that allow balance to be maintained are actions that are obviously advantageous. The adaptive nature of reflex activity will be illustrated by many of the reflexes examined in this exercise.

Reflexes contribute to the maintenance of homeostasis. In **negative feedback mechanisms** a controlled variable is regulated at an appropriate value. The controlled variable is monitored by receptors or sensors that pass information on to an integrator. Deviations of the variable from its ideal level, or **setpoint**, result in activation of effectors that oppose the departure from the setpoint.

A **reflex arc** is a negative feedback circuit in which a detection system is linked to a response system. Reflex action usually requires the following components (Fig. 13-1):

      **\*\* An <u>afferent</u>, or <u>sensory</u>, <u>component</u>. This component includes <u>sensory receptors</u> (these may be modified epithelial cells or afferent neurons) and <u>afferent neurons</u> to transmit impulses to the central nervous system where they synapse with other neurons.**

      **\*\* An <u>integrator</u>, often called an <u>integrating center</u> or a collection of <u>association neurons</u> when it exists within the central nervous system, that determines that magnitude of the response that is appropriate.**

      **\*\* An <u>efferent</u>, or <u>motor</u>, <u>component</u> consisting of <u>neurons</u> that transmit impulses peripherally and an <u>effector</u> that responds to the nerve impulses. The effector often consists of muscle fibers, but it may be a gland.**

**Spinal reflexes** require a functional spinal cord but can occur without brain activity. Although the brain is not required for spinal reflexes to occur, cerebral activity frequently alters spinal reflex action in humans, by either facilitating or inhibiting a reflex. Some of the reflexes to be examined in this exercise require the presence of functional brain tissue if they are to occur. The changes in the size of the pupils of the eyes in response to light are in this category.

**Conditioned reflexes** represent another type of reflex response that requires brain activity. Conditioned reflexes are established by applying two types of stimuli simultaneously, one of which is the appropriate stimulus to elicit a particular reflex response, whereas the other stimulus does not, under most circumstances, cause the response

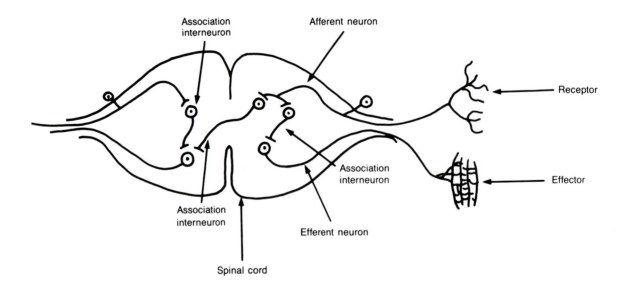

*Figure 13-1. The components of a spinal reflex arc.*

being observed. For example, the sight of food will elicit the secretion of saliva. This secretion is an innate, unconditioned reflex response. If, however, the presentation of food is accompanied by the ringing of a bell, and the two stimuli are given simultaneously many times, the ringing of the bell can eventually cause salivation even though no food is presented. Our behavior patterns involve many conditioned responses to stimuli.

The efferent neurons of a reflex arc may belong either to the somatic or the autonomic motor systems. Somatic neurons innervate skeletal muscle, while autonomic neurons innervate skin, the heart and blood vessels, internal organs, and glands.

In this exercise several reflex responses will be studied. Some of these do not require brain activity, others could not occur without impulses from the brain. The reflexes to be examined will also illustrate the advantages of reflex activity and will demonstrate that several types of tissue can serve as the effector tissue in a reflex response.

In the first part of this exercise, you will demonstrate **stretch reflexes** (also termed **myotatic** or **tendon** reflexes). The receptors for stretch reflexes are called **muscle stretch receptors** or **myotatic organs**. They are embedded among the muscle fibers of almost every skeletal muscle (Fig. 13-2). Stretching a muscle stretches its stretch receptors, resulting in reflexive contraction of the stretched muscle. Stretch reflexes are **monosynaptic**, that is, the afferent neuron synapses directly on the efferent neuron, without an intermediate association neurons (Fig. 13-2).

The **patellar reflex** is an example of a stretch reflex. The patellar reflex can be initiated by tapping the tendon below the **patella**, or kneecap. The stretching of the **extensor muscles**, to which the tendon is attached, results in their reflex contraction and an **extension** (increasing the angle of the joint) of the lower leg. The stimulus we will use to elicit the patellar reflex, a tap on the knee, is an artificial means of stimulating a reflex response. The normal function of stretch reflexes is to oppose involuntary stretching of muscles, making it possible for the body to be held in an erect position and to resist unexpected changes of position. The patella is attached to a large muscle of the upper leg, the **quadriceps femoris**. Contraction

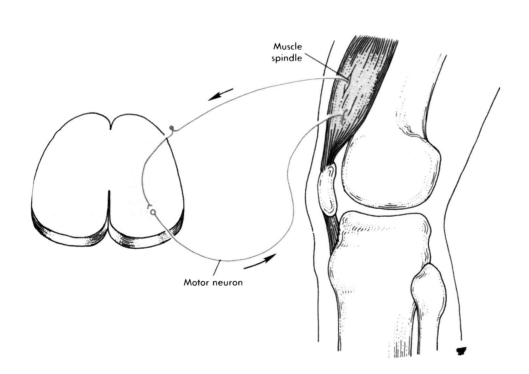

Muscle spindle

Motor neuron

*Figure 13-2. A stretch reflex arc runs from stretch receptors in the stretched muscle to motor neurons that excite the muscle.*

of this muscle extends the leg. When we stand up, the quadriceps femoris is contracted and must generate enough tension to support the weight of the upper body. Stretching of stretch receptors in this muscle occurs if the knee buckles slightly. The resulting stretch reflex opposes this buckling. The **ankle reflex** is another example of a stretch reflex. In this reflex, stretching of the Achilles tendon produces extension of the foot.

The force of a reflex response can be altered if other muscular activity occurs simultaneously. The additional muscular activity reduces the resistance of synapses in reflex arcs in other parts of the body. Mental activity can also influence the expression of reflex activity, in part by increasing muscle tone. On the other hand, fatigue decreases muscle tone. You will observe the effects of mental activity, muscular activity, and fatigue on stretch reflexes in Part 1.

Not all individuals have reflex responses that are equally forceful. It is important, however, that in an individual the reflexes on the two sides of the body be of essentially the same magnitude. A difference in response on the two sides can indicate an abnormality in the nervous system.

In Part 2 of this exercise you will demonstrate reflexes of the eye. In the **pupillary reflex** the effectors are the smooth muscle of the iris, which is innervated by both the sympathetic and parasympathetic branches of the autonomic motor system. The size of the pupil changes in response to changes in input from either branch. Parasympathetic input constricts the pupil, while sympathetic input dilates it. The pupillary reflex opposes changes in the intensity of light reaching the retina; in bright light the pupils constrict, and in dim light they dilate.

A reflex such as the patellar reflex, in which response is seen on the same side of the body to which the stimulus was applied, is said to be **ipsilateral**. A reflex action seen on the side of the body opposite to the side stimulated is called a **contralateral reflex**. Stimulation of one eye by light normally results in both ipsilateral and contralateral reflexes. The constriction of the contralateral pupil is called the **consensual reflex**. This reflex is lost in some conditions in which information transfer from one side of the brain to the other is impaired.

The **ciliospinal reflex** is another reflex in which a change in the size of the pupil is seen. It can be evoked by stimulating the skin on the back of the neck. Stimulating the back of the neck increases the number of sympathetic nerve impulses to muscles that dilate the iris. What function do you suppose the ciliospinal reflex could have?

The first two parts of this exercise deal with reflexes in which the effector tissue is smooth or skeletal muscle. In Part 3 you will see that glands may also serve as effectors and exhibit reflex changes in secretion due to stimulation. The salivary glands are innervated by both branches of the autonomic nervous system. Parasympathetic input favors secretion of large amounts of saliva. Salivary secretion increases reflexively with the taste, smell, or thought of food, in response to mechanical and chemical stimulation of the inside of the mouth, and in response to irritation.

## Procedure

Record your results in the laboratory report section at the end of this exercise.

 *Part 1. Tendon or stretch reflexes*

Work with a partner and take turns being the subject and experimenter.

 *a. Patellar reflex*

1. Subject: sit on a stool or a table with both legs hanging freely.

2. Experimenter: using a percussion hammer, tap the tendon just below the subject's knee. Describe the subject's responses in Table 13-1A.

3. Experimenter: test the other leg in the same way and enter the results in Table 13-1A.

4. Experimenter: write down an addition problem consisting of ten numbers of at least three digits each. Have the subject sit down in a position that allows the patellar reflex to be checked. Check the subject's normal reflex action once or twice, then give the subject the prepared problem and instruct him or her to do it in the shortest time possible without the use of a pencil or pen (or a calculator). While the subject is concentrating on the problem, check his or her patellar reflexes. Describe them in Table 13-1A.

5. Subject: once again assume a position in which the patellar reflexes can be checked.

6. Experimenter: test the patellar reflex. Then instruct the subject to clasp his or her hands together and hold them tightly clasped while contracting the muscles of the arms as if to pull the hands apart. While the subject is engaged in this muscular activity, check the patellar reflexes. Enter the results in Table 13-1A.

7. Subject: run up and down a flight of stairs until you feel quite tired. Then sit down as before.

8. Experimenter: check the subject's patellar reflexes immediately after he or she sits down and compare them to the patellar reflexes observed earlier. Use Table 13-1A to describe the results and compare the strength of the responses observed.

 *b. Ankle reflex*

1. Subject: kneel on a chair or desk, with the ankles protruding beyond the edge of the surface.

2. Experimenter: tap the subject sharply on the Achilles tendon. Record what happens in Table 13-1B.

 *Part 2. Reflexes of the eye*

 *a. Pupillary reflex*

1. Subject: stand in a place where the lighting is relatively dim, away from a window or lamp.

2. Experimenter: note the size of the pupil of each eye and enter this value in Table 13-2A.

3. Experimenter: shine a light into the subject's right eye. Record the diameter of each pupil in Table 13-2A.

 *b. Consensual reflex*

1. Subject: again stand in a place where light is not shining directly into your eyes.

2. Experimenter: note the size of the subject's pupils. Instruct him or her to hold a hand vertically between the two eyes (touching nose and forehead). Shine a light into the subject's left eye. The hand should prevent the light from striking the right eye directly. Record the results in Table 13-2B.

### c. Ciliospinal reflex

1. Experimenter: while observing the subject's pupils, lightly pinch or scratch the back of his or her neck. Describe the response in Table 13-2C.

 *Part 3. Reflexes of the digestive tract*

Both partners should do this part of the exercise.

### a. Salivary reflex

1. Check the pH of your saliva by applying a piece of pH paper to the tip of the tongue. Record the pH in Table 13-3A.

2. Swallow the saliva in your mouth. During the next 5 min collect in a graduated cylinder all the saliva you secrete. Use Table 13-3A to record the volume of saliva you collected.

3. After 5 min put a few drops of vinegar in your mouth. Leave it there for about 5 sec and then spit it in the sink.

4. Check the pH of the your saliva again and collect all the saliva you secrete during the next 5 min. Record this information in Table 13-3A.

5. At the end of the collection period, check the pH of the saliva in your mouth once again. Record this value in Table 13-3A.

### b. Swallowing reflex

1. Swallow the saliva in your mouth.

2. Attempt to swallow again immediately after the first swallow. Record what happened in Table 13-3B.

3. If you were unable to swallow the second time, this might be a) because swallowing is not possible when there is no saliva in the mouth or b) because the swallowing act cannot be performed in rapid succession. To test the second possibility, try rapidly drinking a beaker of water. Describe what happens in Table 13-3B.

### c. Palatal (gag) reflex

1. Touch the soft palate or pharyngeal wall with a tongue depressor. Record what happens in Table 13-3C.

2. Repeat Step 1 while consciously trying to suppress the reflex response. Record the results in Table 13-3C.

*Name:*

*Date:*

*Lab Section:*

## *Analyzing Your Data*

In this exercise you have demonstrated many familiar acts. Use your understanding of how reflex arcs work to explain the observations you recorded in tables 13-1 to 13-3. Be sure you understand the adaptive significance of each of the reflexes you studied.

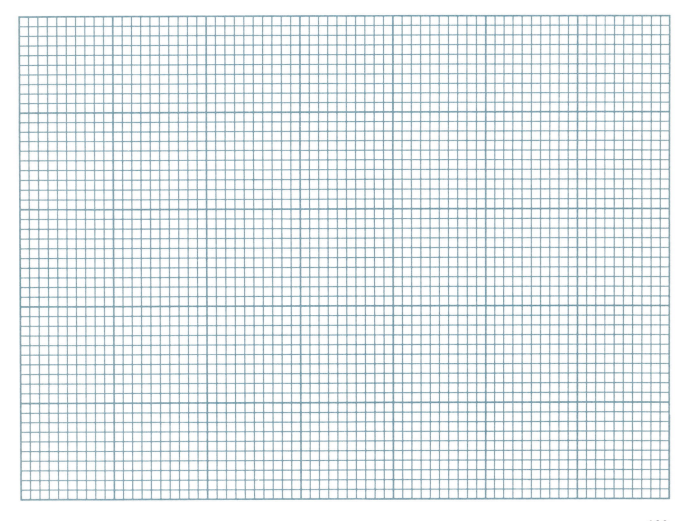

*Table 13-1. Stretch reflexes.*

| A. The patellar reflex. | | |
|---|---|---|
| | Magnitude of response[1] | |
| Conditions under which patellar reflex was tested | Subject 1 | Subject 2 |
| Normal | | |
| Left leg | | |
| Right leg | | |
| During concentration | | |
| Left leg | | |
| Right leg | | |
| During muscular activity | | |
| Left leg | | |
| Right leg | | |
| After strenuous activity | | |
| Left leg | | |
| Right leg | | |
| [1] Weak, moderate, or strong. | | |

*Table 13-1, continued*

| B. The ankle reflex. | | |
|---|---|---|
| | Response | |
| Foot tested | Subject 1 | Subject 2 |
| Left | | |
| Right | | |

*Table 13-2. Reflexes of the eye.*

**A. The pupillary reflex.**

| Experimental conditions | Diameter (mm) | | Change in size | |
| --- | --- | --- | --- | --- |
| | Subject 1 | Subject 2 | Subject 1 | Subject 2 |
| **Left eye** | | | | |
| Dim light | | | | |
| Bright light | | | | |
| **Right eye** | | | | |
| Dim light | | | | |
| Bright light | | | | |

*Pupillary responses*

*Table 13-2, continued*

| | |
|---|---|
| **B. Consensual reflex.** | |
|      **Light not shining directly in eye** | |
| **Left eye** | |
| **Right eye** | |
|      **Light shining only in left eye** | |
| **Left eye** | |
| **Right eye** | |
| **C. Ciliospinal reflex.** | |
|      **Back of neck stimulated** | |
| **Left eye** | |
| **Right eye** | |

*Table 13-3. Reflexes of the digestive tract.*

### A. The salivary reflex.

| | pH of saliva | |
|---|---|---|
| **Experimental conditions** | **Subject 1** | **Subject 2** |
| Start of experiment | | |
| Immediately after vinegar placed in mouth | | |
| 5 min after vinegar placed in mouth | | |
| | Volume of saliva | |
| | **Subject 1** | **Subject 2** |
| 5 min period before vinegar placed in mouth | | |
| 5 min period after vinegar | | |

*Table 13-3, continued*

## B. Swallowing reflex.

| | Ability to swallow | |
|---|---|---|
| Experimental conditions | Subject 1 | Subject 2[1] |
| Start of experiment | | |
| Immediately after swallowing | | |
| While drinking | | |

[1] Yes or no.

## C. Palatal reflex.

| | Response | |
|---|---|---|
| Experimental conditions | Subject 1 | Subject 2 |
| Soft palate touched with tongue depressor | | |
| Soft palate touched while attempting to suppress reflex | | |

## Questions

1. What are the parts of a reflex arc?

2. a) When the subject was concentrating on an arithmetic problem, was the patellar reflex altered? _____

   b) What influence does cerebral activity have on spinal reflexes?

3. Give the receptors and the effector(s) for each of the reflexes listed below:

   a) patellar reflex         _____

   _____

   b) ankle reflex            _____

   _____

   c) pupillary reflex        _____

   _____

   d) consensual reflex       _____

   _____

   e) ciliospinal reflex      _____

   _____

   f) salivary reflex         _____

   _____

g) swallowing reflex  _____

_____

h) palatal reflex  _____

_____

4. What is the function or adaptive advantage of each of the reflexes listed below?

a) patellar reflex  _____

_____

b) pupillary reflex  _____

_____

c) salivary reflex  _____

_____

5. Give one example of an ipsilateral response and one example of a contralateral response.

6. a) What type of innervation to the smooth muscles of the eye causes constriction of the iris? _____

b) What type of innervation of the iris causes dilation?

_____

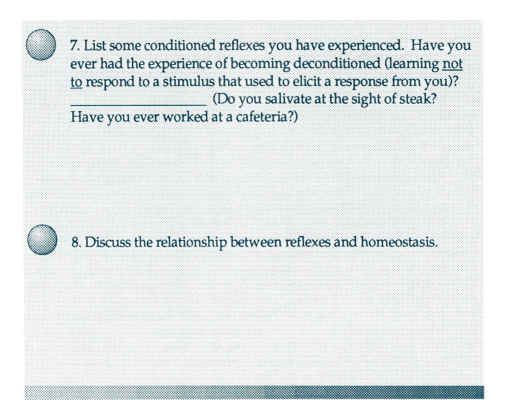

7. List some conditioned reflexes you have experienced. Have you ever had the experience of becoming deconditioned (learning <u>not to</u> respond to a stimulus that used to elicit a response from you)? _____ (Do you salivate at the sight of steak? Have you ever worked at a cafeteria?)

8. Discuss the relationship between reflexes and homeostasis.

# Blood

## Objectives

### Experimental

1. To observe erythrocytes and leucocytes in samples of mammalian and amphibian blood.

2. To observe oxygenated and deoxygenated horse blood.

3. To observe circulation in the arterioles, capillaries, and venules of a frog.

4. To determine students' ABO and Rh blood types.

5. To determine students' hematocrit, hemoglobin content, bleeding time, and coagulation time.

### Conceptual

After completing this exercise and the reading assignment, you should be able to:

1. Compare mammalian and amphibian erythrocytes with respect to a) size, b) shape, and c) presence of a nucleus.

2. List the cellular and fluid components of blood.

3. Name the respiratory pigment of vertebrates and describe its structure and the color changes associated with reversible binding to oxygen.

4. Define the terms **arteriole, venule, antibody, antigen, agglutination,** and **hemostasis**.

5. Determine the ABO and Rh blood type of a sample if its agglutination reactions with anti-A, anti-B, and anti-Rh sera are known.

6. Describe the circumstances that can lead to **erythroblastosis fetalis**.

7. Discuss the clinical significance of **hematocrit, hemoglobin content, bleeding time,** and **coagulation time** and explain how each of these is measured.

8. Summarize the events that occur in coagulation.

## Background

Blood is made up of cells suspended in liquid **plasma**. **Erythrocytes** (red blood cells [Fig. 14-1]), **leucocytes** (white blood cells [Fig. 14-2]), and **platelets** (cell fragments involved in blood clotting) comprise the cellular elements in blood. In Part 1 of this exercise, you will observe erythrocytes and leucocytes in amphibian and mammalian blood.

Vertebrate blood contains the respiratory pigment **hemoglobin**, which binds reversibly with $O_2$. A molecule of hemoglobin contains four polypeptide chains, each of which contains a **heme** group, an iron-containing portion capable of binding $O_2$. Hemoglobin that is saturated with $O_2$ is bright red; deoxygenated hemoglobin is dark red. You will observe this phenomenon in the demonstration of oxygenated and deoxygenated blood in Part 2.

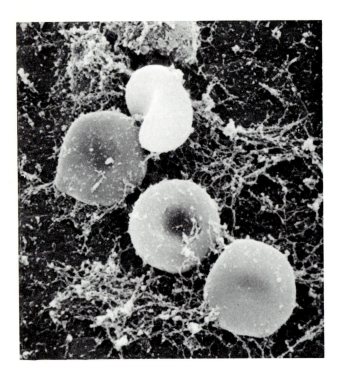

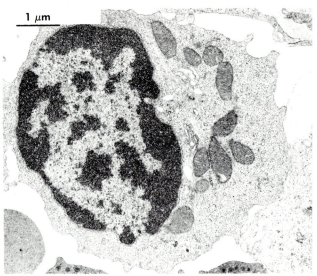

*Figure 14-2. A transmission electron micrograph of a white blood cell.*

*Figure 14-1. A scanning electron micrograph of red blood cells.*

Exchange of nutrients, $O_2$, and wastes takes place in the smallest blood vessels, the **capillaries**. Arteries carrying oxygenated blood branch to give rise to **arterioles**, and arterioles branch to form capillaries. Capillaries merge to give rise to the smallest veins, or **venules**. In Part 3 you will examine a demonstration of microcirculation in a frog. When you examine the circulating blood, venous blood will not appear darker than arteriolar blood. However, you should be able to distinguish arterioles from venules, because a) only the blood in the arterioles pulsates and b) in arterioles the blood travels from a larger vessel to vessels with smaller diameters, whereas in venules blood from smaller vessels merges into larger vessels.

**Antibodies** are plasma proteins that recognize foreign substances and activate the immune system. An **antigen** is a macromolecule that produces a specific immune response. Erythrocytes possess surface antigens that lead to **transfusion reactions**, a type of tissue graft rejection in which blood is the tissue that is transplanted. If a person's blood contains antibodies to foreign antigens that are received during a transfusion, the antibodies will cause clumping (**agglutination**) of the transfused cells and destroy them.

One example of such an antigen is **Rh factor**. A person with Rh antigens is $Rh^+$, someone lacking these antigens is $Rh^-$. **A person who is $Rh^-$ possesses antibodies to Rh antigens only if he or she has been exposed to these antigens.** This can happen during a transfusion or when an $Rh^-$ mother gives birth to an $Rh^+$ infant. During most of gestation, fetal red blood cells do not normally enter the mother's circulation, but they may do so at birth. If a mother who is $Rh^-$ gives birth to an $Rh^+$ infant, the mother will acquire anti-Rh antibodies from exposure to the Rh antigen. In a subsequent pregnancy with an $Rh^+$ fetus, the mother's antibodies will attack the fetus, causing **Rh disease** or **erythroblastosis fetalis** (Fig. 14-3).

The most familiar family of red blood cell antigens is the basis for the **ABO blood group system**. A person's ABO blood type is determined by the antigens present on his or her erythrocytes. If you have A antigens on your erythrocytes, you have type A blood, and if you have B antigens, you have type B blood. If you have both A and B, your

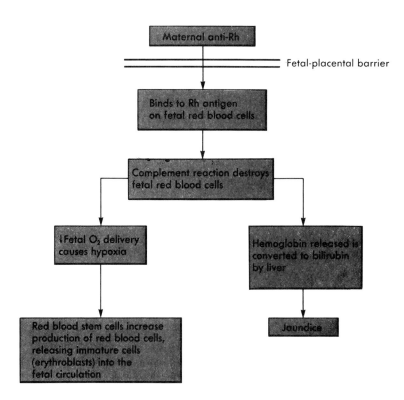

*Figure 14-3. Steps in the development of Rh disease.*

blood type is AB; if you have neither A nor B antigens, you have type O blood (see Table 14-1). The ABO blood type system is unusual because **a person has antibodies in his or her blood plasma for the antigens he or she does not possess, even without prior exposure to those antigens**. If you receive a transfusion of blood containing foreign antigens in the ABO group, a transfusion reaction will occur, even if you have not encountered the foreign antigen before.

In Part 4 of this exercise, you will have an opportunity to determine your blood type with respect to the ABO system and Rh factor. To determine student blood types, anti-A and anti-B sera will be used. Anti-A serum contains antibodies against A antigens; anti-B serum contains antibodies against B antigens (Table 14-1).

There are actually many antigens and incompatible plasma antibodies in addition to those of the ABO and Rh systems. For this reason the blood of donor and recipient should be mixed, to determine whether agglutination occurs, before a transfusion is given.

Parts 5, 6, and 7 of this exercise deal with the oxygen-carrying abilities of blood. The oxygen-carrying capacity of blood can be assessed by determining the percentage of blood that is made up of red blood cells (**hematocrit**) or by measuring the **hemoglobin content** of the blood. **Anemia** occurs when one or more of these measurements if abnormally low.

**Centrifugation** speeds up the settling process and is used clinically to separate small samples of whole blood into a plasma fraction and a cellular fraction, so that hematocrit can be determined (Fig. 14-4). Normally, in adult males about 47% of the volume of whole blood is cells, and in adult females hematocrit is about 42%. Almost all of this is red cells. The white cells and platelets form a small layer on top of the red cells.

Since hemoglobin is the major constituent of red cells, hematocrit is a useful index of the oxygen-carrying ability of blood. However, sometimes it is desirable to determine hemoglobin concentration rather than hematocrit. The **Tallquist scale** is used to estimate g of hemoglobin per 100 ml of

*Table 14-1. Characteristics of blood types A, B, AB, and O.*

| Blood type | Antigens present | Antibodies present | Can accept from | Can donate to |
|---|---|---|---|---|
| A | A | anti-B | A or O | A or AB |
| B | B | anti-A | B or O | B or AB |
| AB | A and B | none | A, B, AB, O | AB |
| O | none | anti-A, anti-B | O | A, B, AB, O |

| If agglutination occurs in: | The blood type is: |
|---|---|
| Anti-A serum only | A |
| Anti-B serum only | B |
| Anti-A and Anti-B | AB |
| Neither | O |

blood. In this method the color of a drop of blood is compared to a standard color chart. This is an easy, but not very reliable, method of measuring hemoglobin. You will use this technique in Part 5.

When a person receives a cut or scrape, blood loss is normally limited by several processes collectively known as **hemostasis**. There are three stages of hemostasis (Fig. 14-5): a) the sealing of the damaged vessel by a temporary **platelet plug**, b) the formation of a **clot** (Fig. 14-6), a network of threadlike **fibrin** molecules that forms a more lasting patch over the damaged vessels, and c) the sealing of the clot after the vessels have been repaired.

Measures of the speed with which a wound stops bleeding (Part 6) and a clot forms (Part 7) are clinically useful. **Bleeding time** is taken as the time for a small sharp incision to stop bleeding. Normal bleeding time ranges from 1 to 3 min. However, the time depends upon the depth of the wound and the amount of blood circulation to the injured part as well as upon the presence of adequate amounts of clotting factors.

**Coagulation** (or clotting) **time** is the time required for a sample of blood removed from the body to clot. The usual way to determine coagulation time is to allow blood to flow from an incision

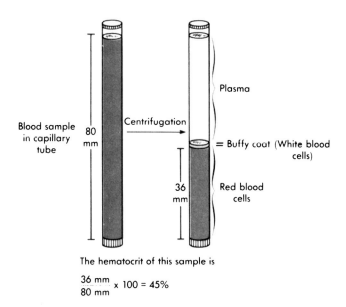

*Figure 14-4. Separation of blood by centrifugation into cellular and fluid (plasma) components. A blood sample is placed in a glass capillary tube treated to prevent clotting. After centrifugation, the red cells are packed at the bottom of the tube with the white cells above them; the plasma remains at the top of the tube.*

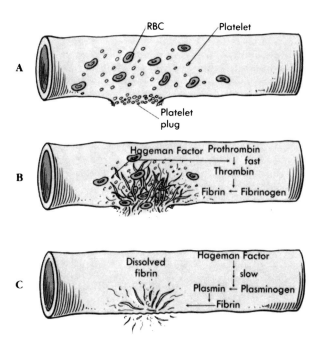

*Figure 14-5. The sequence of events in formation and subsequent dissolution of a blood clot. A A platelet plug is formed. B Damage activates Hageman factor, which initiates a cascade of reactions (**arrows**) that ends with formation of a blood clot. Red and white blood cells and platelets are trapped among the fibrin strands of the clot. C The clot is eventually dissolved by plasmin which is produced by a much slower reaction sequence also triggered by Hageman factor.*

*Figure 14-6. A scanning electron micrograph of a blood clot, showing trapped cells.*

into a glass capillary tube. The tube is broken at 30 sec intervals to determine when clot formation occurs. Normal clotting time ranges from 2-6 min. However, the diameter of the capillary tube influences coagulation time.

## *Procedure*

Record your results in the laboratory report section at the end of this exercise.

 *Part 1. Blood cells*

1. Obtain a drop of frog blood that has been diluted 1:5 with 0.7% NaCl. This solution has the same NaCl concentration as frog blood plasma. The blood suspension should have a pale red color.

2. Place a drop of the blood suspension on a microscope slide, cover it with a cover slip, and examine under low magnification. If the mixture is too dense to permit visualization of individual cells, use the high power lens.

3. Note the size, shape, and color of the erythrocytes and whether nuclei are present. Record this information in Table 14-2.

4. Look for leucocytes. These are relatively rare and difficult to see, so don't give up right away. If you find one, sketch its shape and color and note whether a nucleus is present (Table 14-2). Inform your instructor of your discovery so others can take a look.

5. Repeat steps 1 to 4 with a sample of mammalian blood diluted with 0.9% NaCl. Compare the red blood cells of the mammal and the frog with respect to size and shape. Record your observations in Table 14-2.

 *Part 2. Oxygenated and deoxygenated blood*

1. Compare the color of a small sample of horse blood which has had oxygen bubbled through it with a similar sample through which nitrogen has been bubbled.

 *Part 3. Capillary circulation*

1. Observe the demonstration of circulation in arterioles, capillaries, and venules of an anesthetized frog. The tongue or the webbing of the hind foot will be stretched over a piece of balsa wood or cardboard with a hole beneath the stretched skin. Use a dissecting microscope to observe circulation in the skin.

Before proceeding with the rest of this exercise, read the following statement carefully.

CAUTION: As you probably know, Acquired Immune Deficiency Syndrome (AIDS), as well as other diseases, can be transmitted through infected blood or blood products. Since many of you will be entering the health professions, it is essential that you learn the proper procedures for handling blood samples to prevent disease transmission. It is extremely important that all students, instructors, and staff observe the following precautions:

** All human blood samples and antisera should be treated as potentially contaminated.

** Blood samples and antisera should never be touched or handled directly.

** Any person handling used lancets, capillary tubes, slides, empty serum bottles, droppers, toothpicks, tissues, filter paper, pins, and other materials that have been in contact with blood or serum should wear disposable, latex gloves and should wash his or her hands after removing the gloves.

** Only sterile lancets and clean, unused capillary tubes, slides, toothpicks, tissues, filter paper, pins, and microscope slides should be used. NEVER REUSE LANCETS!!!!!

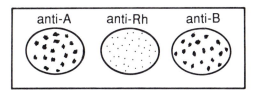

*Figure 14-7. Slide with antisera for determining ABO and Rh factor blood types. Agglutination has occurred in the anti-A serum and the anti-Rh serum.*

** All used lancets, capillary tubes, and slides should be placed in a 10% bleach solution immediately after use.

** All gloves, empty serum bottles, droppers, toothpicks, tissues, filter paper, pins, and other materials that have been in contact with blood or serum should be placed in sealed, double plastic bags and disposed of in the proper containers.

** All waste receptacles containing blood products should be labeled accordingly.

Parts 4 and 5 should be done individually. For parts 6, 7, and 8 students should work in groups of four, with each group member doing either Part 6, 7, 8a, or 8b.

 *Part 4. Blood typing*

1. Using a wax pencil, draw three complete circles on a glass slide. The circles will prevent blood-antisera mixtures from running together.

2. Write the letter *A* under the circle on the left, write *B* under the middle circle, and write *Rh* under the circle on the right (Fig. 14-7).

3. Obtain a capillary tube for hematocrit determination and have it ready. The inner surface of the tube has been heparinized to prevent clotting.

4. Clean a finger with 70% alcohol and let it dry.

5. Puncture your fingertip with a sterile lancet. Collect the first few drops of blood in the hematocrit tube and set it aside. The blood will enter the tube by capillarity if you just touch the end of the tube to the drop on your finger. Dispose of the lancet properly.

6. Place one drop of blood in each of the circles on your slide.

7. Place one drop of anti-A serum in the circle on the left, one drop of anti-B serum in the middle circle, and one drop of anti-Rh serum in the circle on the right. (Do not allow the serum to come in contact with your skin, especially not with the cut on your finger.)

8. Mix the blood with the antiserum by means of a toothpick. To avoid contamination use a separate toothpick for each stirring operation. Dispose of the toothpicks properly.

9. Tip the slide carefully back and forth for one minute. (Do not let the samples run together.)

10. Examine the slide for evidence of agglutination. This should begin almost immediately and should be complete by the end of 1 min. Agglutination of the red cells will give a speckled appearance to the field (Fig. 14-7). Each speck is a focus of agglutination. A nonagglutinated field will have a uniform, murky appearance. Consult your instructor if you are in doubt.

11. Determine your blood type (see Table 14-1) and record your results on the board. Determine the percentages of each blood type in the class population (Table 14-3).

12. Dispose of slides and lancets properly.

*Part 5. Determining hematocrit*

1. Cap your hematocrit tube with the clay provided for this purpose.

2. Place it in the centrifuge and note the number of the slot you put it in (Table 14-4).

3. After the tubes have been centrifuged, measure the length of tube which contains red cells and the total blood volume (cells plus plasma). Enter these values in Table 14-4.

4. Calculate the hematocrit of your sample.

5. Dispose of capillary tubes properly.

 *Part 6. Determination of the hemoglobin content of blood*

1. Remove a small piece of absorbent tissue from the Tallquist test booklet.

2. Clean your finger with 70% alcohol. Let the alcohol evaporate.

3. Puncture the finger with a sterile lancet. Dispose of the lancet properly.

4. Wipe away the first drop of blood with the edge of a Tallquist tissue.

5. Place the second drop of blood in the center of the tissue. (<u>Do not squeeze the finger</u>. The drop must flow freely.)

6. In a few seconds the blood stain will lose its glossiness. At that time the blood should immediately be compared with the Tallquist color chart under direct, not artificial, light. <u>Do not let the blood dry to a brown color</u>, or inaccurate values will be obtained.

7. Place the specimen under the Tallquist color chart and move it so the blood stain appears under the apertures.

8. Locate the color on the Tallquist scale that most closely matches the color of the blood stain. Since the color differences on the chart represent 10% variations in hemoglobin content, it will be necessary to estimate the percentage of hemoglobin. Enter your results in Table 14-5.

9. Dispose of all tissues properly.

10. Enter your results on the board and calculate the class averages for males and females. Enter this information in Table 14-5.

 *Part 7. Bleeding time*

1. Puncture your finger as directed earlier.

2. Induce a free flow of blood. Collect each drop of blood that appears on a piece of filter paper. The first drop on the filter paper should be 1 cm or more in diameter.

3. Blot the blood with filter paper every 30 sec until the bleeding stops. Enter the number of drops on the paper in Table 14-6.

4. Divide the number of drops on the filter paper by 2. This gives the bleeding time in min. Record this value (Table 14-6).

5. Dispose of lancets and filter paper properly.

6. Enter your results on the board and calculate the class average. Enter this in Table 14-6.

 *Part 8. Coagulation time*

 *a. Slide method*

1. Clean your finger with 70% alcohol and allow it to dry.

2. Puncture the finger tip with a sterile lancet. Do not use the first drop of blood, since its clotting time is abnormally low.

3. <u>Note the time</u> that the drop of blood to be used first appears in Table 14-7. This time is used as the beginning of the experiment.

4. Place the drop of blood on a glass slide.

5. At 30 sec intervals draw a straight pin <u>slowly</u> through the blood and observe the point of the pin.

6. Repeat Step 5 until fine red threads can be detected on the end of the pin. This is the end point. Sometimes the entire mass forms a gel. In that case, the time of gel formation is considered the end point.

7. The time from the appearance of the drop of blood until the appearance of the first thread is the coagulation time. Using this method of determination, the normal range of coagulation times is from 2 to 8 min. Record your coagulation time in Table 14-7.

8. Dispose of lancets, pins, and slides properly.

9. Enter your results on the board and calculate the class average (Table 14-7).

### b. Capillary tube method

1. You will need a large drop of blood for this test. Sterilize the tip of your ring finger with 70% alcohol. Swing the arm a few times to increase blood flow to the hand; then puncture the fingertip with a <u>firm</u> jab. If necessary, press gently to stop the flow. Wipe away the first drop and use only freely flowing blood.

2. In Table 14-7 note the time the drop of blood to be used appears. This time is used as the beginning of the experiment.

3. Place one end of the capillary tube into the drop of blood. Keep the other end of the tube open. Hold the tube horizontally (do not let the tip touch the skin) and allow the tube to fill by capillarity.

4. After exactly 1 min carefully break off about 1 cm of the capillary tube and determine whether a thread of coagulated blood is visible between the two pieces of tubing.

5. Repeat Step 4 every 30 sec until such a thread is obtained.

6. Record the coagulation time in Table 14-6.

7. Dispose of lancets, capillary tubes, and tube fragments properly.

8. Enter your results on the board and calculate the class average (Table 14-7).

# Laboratory Report

## *Exercise 14: Blood*

Name: _____

Date: _____

Lab Section: _____

## Analyzing Your Data

Review the data recorded in the tables before
answering the questions that follow.

*Table 14-2. Characteristics of erythrocytes and leucocytes.*

| Type of cell | Size | Shape | Presence of Nuclei |
|---|---|---|---|
| **A. Frog blood** | | | |
| Erythrocytes | | | |
| Leucocytes | | | |
| **B. Mammalian blood** | | | |
| Erythrocytes | | | |
| Leucocytes | | | |

**Table 14-3.** *Class frequencies of ABO and Rh blood types.*

| Blood type[1] | Number with this blood type | Number tested | Percentage |
|---|---|---|---|
| A$^+$ | | | |
| B$^+$ | | | |
| AB$^+$ | | | |
| O$^+$ | | | |
| A$^-$ | | | |
| B$^-$ | | | |
| AB$^-$ | | | |
| O$^-$ | | | |

[1] Letters refer to ABO blood type; + and - refer to Rh factor.

**Table 14-4.** *Hematocrit of student blood samples.*

| | |
|---|---|
| Centrifuge tube number: | |
| Length of erythrocyte column (cm): | |
| Length of blood column in tube (cm): | |
| Hematocrit = $\dfrac{\text{Length of erythrocyte column} \times 100}{\text{Length of blood column}}$ = | |
| Class average for males: | |
| Class average for females: | |

*Table 14-5. Hemoglobin content of student blood samples.*

| |
|---|
| Estimated hemoglobin content of your blood: |
| Class average for males: |
| Class average for females: |

*Table 14-6. Bleeding time of student blood samples.*

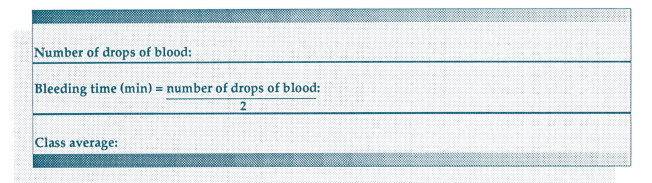

| |
|---|
| Number of drops of blood: |
| Bleeding time (min) = number of drops of blood: $\frac{}{2}$ |
| Class average: |

*Table 14-7. Coagulation time of student blood samples.*

**A. Slide method**

Starting time:

Ending time:

Total time (min):

Coagulation time (min):

Class average:

**B. Capillary tube method**

Starting time:

Ending time:

Total time (min):

Coagulation time (min):

Class average:

# *Questions*

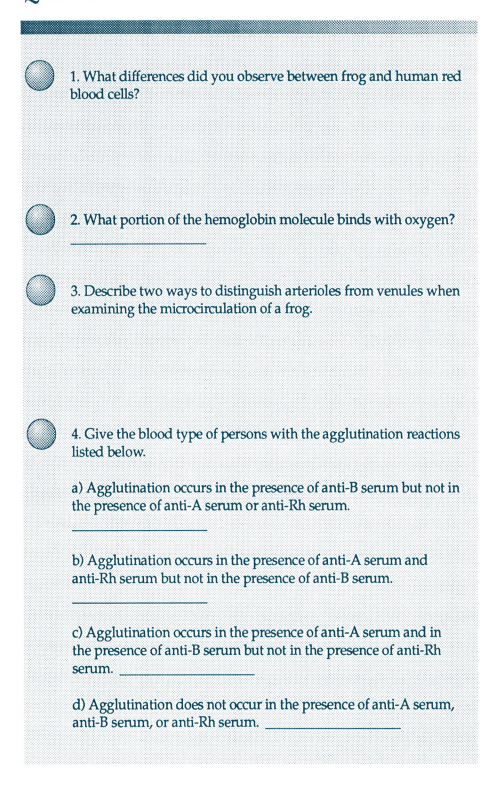

1. What differences did you observe between frog and human red blood cells?

2. What portion of the hemoglobin molecule binds with oxygen?

_____

3. Describe two ways to distinguish arterioles from venules when examining the microcirculation of a frog.

4. Give the blood type of persons with the agglutination reactions listed below.

a) Agglutination occurs in the presence of anti-B serum but not in the presence of anti-A serum or anti-Rh serum.

_____

b) Agglutination occurs in the presence of anti-A serum and anti-Rh serum but not in the presence of anti-B serum.

_____

c) Agglutination occurs in the presence of anti-A serum and in the presence of anti-B serum but not in the presence of anti-Rh serum. _____

d) Agglutination does not occur in the presence of anti-A serum, anti-B serum, or anti-Rh serum. _____

5. Specify what antibodies of the ABO system will be present in the serum of a person with each of the blood types listed below:

a) Type A: _____

b) Type B: _____

c) Type AB: _____

d) Type O: _____

6. Under what circumstances might an agglutination reaction involving Rh antigens and antibodies occur?

7. What is the major difference between the immune responses of the Rh system and the ABO system?

8. How is hemostasis maintained?

# 15

# *Regulation of Heart Rate*

*Reading assignment: text 275-280*

## *Objectives*

 *Experimental*

1. To determine the effect of stimulation of the vagus nerve on heart rate in the turtle.

2. To determine the effects of application of a) acetylcholine, b) atropine, and c) epinephrine on heart rate in a turtle.

3. To determine in which part of the turtle heart the beat originates.

 *Conceptual*

After completing this exercise and the reading assignment, you should be able to:

1. State the structure responsible for pacemaker activity in a) reptiles and b) mammals.

2. Describe the effects of sympathetic and parasympathetic inputs on heart rate and name the nerves responsible for these inputs.

3. Define the terms **cholinergic, muscarinic, nicotinic, agonist**, and **antagonist**.

4. Describe the effect of stimulation of the vagus nerve on the heart, the mechanism by which this effect is produced, and the evidence for this mechanism.

5. Describe the effects of a) acetylcholine, b) atropine, and c) epinephrine on the heart and the mechanism by which this effect is produced.

## *Background*

Excitation of the vertebrate heart is myogenic, that is, contraction of the heart originates within the muscle itself. In reptiles such as the turtle you will use in this exercise, the pacemaker is a structure called the **conus arteriosus**; in mammals the **sinoatrial**, or **SA, node** is the pacemake (Fig. 15-1). The frequency of depolarization of the pacemaker is modified by inputs from the autonomic nervous system. Parasympathetic input occurs via the **vagus** nerve (cranial nerve X) and slows the heart rate. Sympathetic input occurs via the **cardiac sympathetic** nerves and increases the heart rate

(Fig. 15-2). Note that these nerves are the efferent pathways that are responsible for control of intestinal motility and arterial blood pressure.

In parasympathetic pathways the final motor neuron that synapses directly on the target organ is almost always located in a ganglion or plexus that is right in or on the target organ. The efferent vagus fibers synapse on the cells of these plexuses or ganglia. The transmitter at the synapse between the vagus fiber and the ganglion or plexus cells is **acetylcholine**, which is also the transmitter at the synapse between ganglion cells and the target organ.

*Figure 15-1. Location of the sinoatrial node on the right atrium of the mammalian heart.*

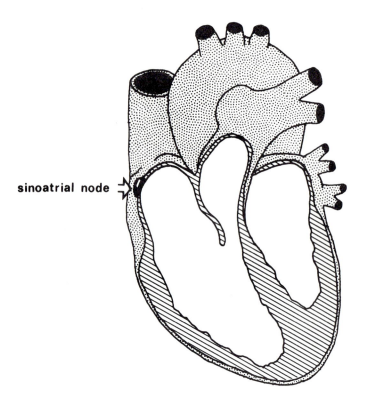

sinoatrial node →

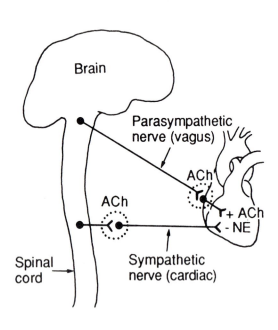

*Figure 15-2. Autonomic innervation of the heart. Dotted lines indicate ganglia. ACh indicates acetylcholine; NE indicates norepinephrine.*

The sympathetic fibers that innervate the heart run from sympathetic ganglia located close to the spinal cord. These fibers synapse directly on the heart and release the neurotransmitter **norepinephrine**. In addition, **epinephrine** (along with some norepinephrine) reaches the heart from the adrenal medullae via the blood.

Organs that are innervated by the autonomic nervous system have specific receptors for acetylcholine and epinephrine. Receptors for acetylcholine are **cholinergic**. There are two types of cholinergic receptors: **muscarinic** and **nicotinic** (Fig. 15-3). Muscarinic receptors are cholinergic receptors located on effectors receiving parasympathetic innervation.

Drugs or toxins that activate a particular receptor type are **agonists**, while those that interfere with the function of a receptor are **antagonists** (Fig. 15-4). **Atropine** is a naturally occurring plant compound that acts by binding to acetylcholine receptor sites of effector organs and blocking

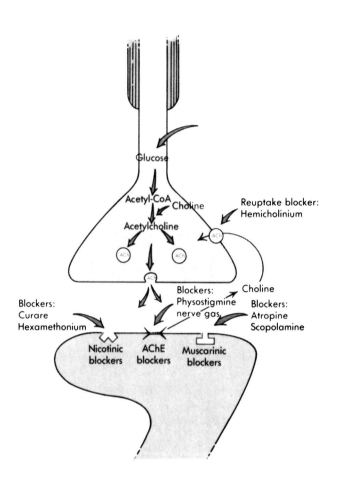

*Figure 15-3. Neurotransmission at cholinergic synapses. Both nicotinic and muscarinic receptor types are shown together for simplicity, although actually both are not found together at the same synapse. Cholinergic transmission can be blocked by receptor antagonists and by inhibitors of choline absorption. Transmission can be enhanced by inhibitors of acetylcholinesterase.*

them so that they cannot be reached by acetylcholine molecules. Thus, atropine is a muscarinic antagonist (Fig. 15-3).

In this exercise, you will study the effects of stimulating the vagus nerve on heart rate. To do this, you will address the following sequence of questions:

**\*\* What effect do impulses from the vagus nerve have on the heart?** (See Part 2: steps 2 to 4, below.)

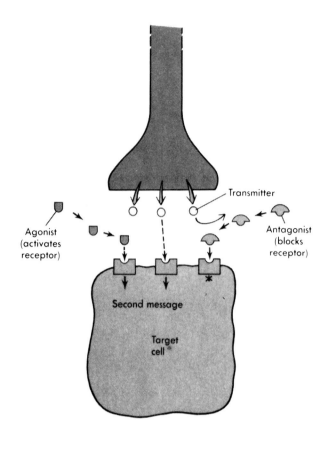

*Figure 15-4. Agonists bind to postsynaptic receptors and have an effect similar to that of neurotransmitters. Antagonists block the receptor, interfering with transmission.*

**\*\* What structure of the heart is affected by vagal input, the pacemaker or the heart muscle?** (See Part 2: Step 3, below.) (If you cannot determine this directly, can you rule out any structures?)

**\*\* How do impulses from the vagus nerve produce their effects?** To answer this question, you will test a hypothesis formulated from what you have learned by reading your text.

**Hypothesis 1:** Acetylcholine released by nerve endings in the vagus nerve produces the observed change in heart rate.

Steps 8 to 15 of Part 2 will allow you to gather the information you will need to test this hypothesis.

To be sure you understand all parts of this experiment, complete the predictions below.

**Prediction 1:** Adding acetylcholine to the heart will _____.

**Prediction 2:** Adding acetylcholine plus atropine to the heart will

_____.

**Prediction 3:** Adding acetylcholine plus atropine to the heart and simultaneously stimulating the vagus nerve will

_____.

Because a scientist doesn't know beforehand whether or not a hypothesis is correct, it is a good idea to design some experiments to test alternative hypotheses as well. In this exercise, we will investigate an alternative hypothesis involving a different mechanism.

**Alternative hypothesis:** Epinephrine produces the observed change in heart rate.

Directly stimulating the sympathetic nerves to the heart is rather difficult. Therefore, to imitate the action of the sympathetic nerves, you will add epinephrine to the heart.

## Procedure

Record your results in the laboratory report section at the end of this exercise.

 *Part 1. Turtle dissection*

1. The experimental animal we will use for this exercise will be the turtle. Turtles harbor <u>Salmonella,</u> a variety of bacteria that causes unpleasant intestinal upsets. <u>Keep your hands away from your mouth, nose, and eyes and wash your hands thoroughly at the end of the period.</u>

2. You will be supplied with a single-pithed turtle. In addition, the bony shell bridge connecting the <u>carapace</u> (the dorsal part of the shell) and the <u>plastron</u> (the ventral shell) will have been sawed through.

3. Although the brain of the turtle has been destroyed, the spinal cord is intact. Consequently, the limbs can move around and will interfere with measurements unless they are immobilized. To do this, place the turtle on its back on the board. Tie a length of string around each leg and fasten the strings to the board by tying through the holes at the corners. Similarly, pull the head out so that the neck is extended and fasten it to the central hole at one end of the board. Your turtle should now be essentially immobile, and there should be no wrinkles in the neck skin (Fig. 15-5).

4. Carefully cut the skin at the head and tail ends of the plastron and, raising the edge of the bone, cut against the plastron to free it completely. The heart may now be seen through its connective tissue covering.

5. In what follows, use only <u>blunt</u> instruments (forceps, scissors, blunt probes). <u>Do not use a scalpel.</u> Cut through the skin, and only the skin, from the jaw to the beginning of the plastron. Make a slit about 2 cm long. Be careful to avoid cutting blood vessels; excessive bleeding will impair the function of both the heart and the nerve. Ask your instructor for help if you have trouble at this point.

6. Cut laterally at the anterior end of this slit (under the jaw) and fold the skin back.

7. Using your probe or the closed tips of your scissors as tearing instruments, split the muscle layer on one side of the trachea and expose the <u>carotid artery</u>. (This will be obvious, because it is dark red, unless the turtle has lost too much blood). Don't do any cutting during this process, since you may cut blood vessels or the vagus nerve. The vagus is a white fiber running along the carotid artery in a sheath of connective tissue.

8. Carefully open the sheath, expose a stretch of nerve, and pass a thread moistened with turtle Ringer's solution under it. This thread is called a <u>ligature</u>. Don't knot the thread around the nerve. The ligature can be used to lift the nerve onto the stimulating electrodes later.

9. Keep the nerve and the heart moist with plenty of Ringer's solution throughout the experiment.

10. Open the <u>pleura</u> (the sheet of epithelium lining the thoracic cavity) and the <u>pericardium</u> (the fibrous sac enclosing the heart), to expose the heart. Take time to carefully observe the cycle of contraction of atria and ventricle. When the ventricle contracts blood is prevented from entering the blood vessels that serve the ventricle. During relaxation the ventricular blood vessels are filled with blood. This is why the ventricle is red when it is relaxing and pale when it is contracting.

11. Note that the tip of the ventricle connects to the pericardium. The connection is called the <u>frenulum</u>. Cut the frenulum free from the pericardium and attach the hook of your heart lever to the frenulum, not through the ventricle. Attach a thread from the hook to the lever. You can now record each beat on the recorder (Fig. 15-5).

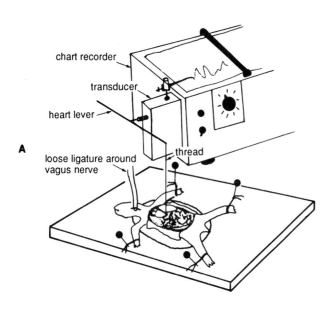

A

chart recorder

transducer

heart lever

loose ligature around
vagus nerve

thread

B

*Figure 15-5. **A** Turtle heart dissection. **B** A stimulator.*

**12. Choose a paper speed that clearly shows the shape of individual beats (try 0.25 cm/sec) and obtain a "resting" heart beat. Keep the heart moist with turtle Ringer's.**

 *Part 2. Making chart recordings*

It is essential that you note on the recording every-thing you do: when you stimulate the nerve, when you stop stimulating, when you add or remove a solution, and so forth. Record paper speed, time, and all treatments that were applied to the heart as well as your observations about qualitative changes in heart beat (such as strength of contrac-tion) and anything else you noticed.

**1. Record about 2 min of normal heart activity. Because you know the paper speed, you can calculate heart rate as follows:**

Paper speed (cm/sec) x 60 sec/min = paper speed
in cm/min.

When you know the distance the paper moves each minute, count the number of heartbeats that occur in that many cm to get the heart rate in beats/min. Your results will be more accurate if you count the number of beats occurring during several minutes and divide by the number of minutes.

**2. Gently lift the vagus nerve slightly with the thread and slip the two prongs of the stimulating electrode under it. Stimulate the vagus. Begin with an intensity of about 1 V and a frequency of 60/sec. If the heart stops completely, go on to Step 3. If no change occurs, reverse the polarity of the stimulating electrode. If this is not effec-tive, increase the voltage by successive steps of 3 V until the heart stops. If this does not work, ask for help.**

**3. Continue stimulating the vagus nerve. With the heart stopped, poke the ventricle lightly with the sharp tip of a dissecting needle. <u>Think about whether stimulation of the vagus nerve affected the ability of the cardiac muscle to contract</u>.**

**4. Continue the stimulation and wait. Ultimately contraction will begin. <u>Notice which part of the heart contracts</u>.**

5. Stop stimulating the nerve. <u>Observe whether the first few beats of the heart are stronger than, weaker than, or of the same magnitude as the normal beats you observed in Step 1</u>.

6. Allow the heart to rest for a few minutes, until it returns to a rate similar to that you observed before vagal stimulation.

7. Wash the heart thoroughly with Ringer's solution; then use a dropper to remove as much of the solution as possible from around the heart. Record another 5 cm of normal heartbeat.

8. Add 3 to 4 dropperfuls of acetylcholine (0.06 M) to the atria, especially around the right side. (Remember where the pacemaker is located.)

9. Allow the heart to recover as before.

10. Remove as much of the acetylcholine solution as possible with a dropper and replace it with Ringer's. Record 5 cm or more of normal heartbeat.

11. Stimulate the vagus nerve again, just enough to show that it's still working. With the heart beating normally, add 3 dropperfuls of atropine (0.05 M) around the atria.

12. Observe the effects of the atropine and wait for a steady heart rate. Without removing the atropine, add the same amount of acetylcholine as in Step 8.

13. Without changing the solutions, stimulate the vagus as you did before. <u>Observe how the effects on the heart compare with the last time you stimulated the vagus</u>.

14. Remove the liquid surrounding the heart with a dropper and apply fresh Ringer's solution.

15. Remove the Ringer's and apply several drops of epinephrine (0.01 M) to the fluid surrounding the heart. If the first application of epinephrine is ineffective, drain the pericardial puddles with your dropper and add more epinephrine.

16. Once you have seen a clear effect of epinephrine, rinse the heart with fresh Ringer's and wait for it to return to a resting rate.

17. Now cut the heart loose from its veins, arteries, connective tissue and the heart lever. Put the heart in a beaker of Ringer's.

18. Carefully separate the two atria from the ventricle with scissors. Replace the heart pieces in fresh Ringer's. <u>Notice which part(s) continue(s) to beat</u>.

# Laboratory Report

*Exercise 15: Regulation of Heart Rate*

Name: _____

Date: _____

Lab Section: _____

## Analyzing Your Data

Be sure you understand how each of the steps of the experimental procedure relates to the hypotheses formulated at the beginning of this exercise.

As in earlier exercises, in this experiment your chart recordings provide a record of your data. Be sure each diagram is clearly labeled and contains complete information including recorder speed, time, and all treatments that were applied to the heart.

Before you leave, go over your chart recordings to be sure you understand them. Make sure that you have clearly indicated everything you did on the recordings, so that you will be able to interpret them later. Use Table 15-1 to summarize your results. Using your knowledge of physiology, your chart recordings, and your data summary, you should be able to determine the effects of autonomic stimulation on the heart and the mechanism by which this effect is produced.

*Table 15-1. Summary of effects of vagal stimulation and application of acetylcholine, atropine, and epinephrine on turtle heart rate.*

| Paper speed: | | |
| --- | --- | --- |
| | | **Responses** |
| **Condition or Treatment** | **Heart rate (beats/min)** | **Remarks[2]** |
| Control | | |
| Vagus nerve stimulated | | |
| Heart prodded during vagal stimulation | | |
| Vagal stimulation continued | | |
| Vagal stimulation stopped | | |
| Control[1] | | |
| Acetylcholine applied | | |
| Control[1] | | |
| Atropine applied | | |
| Acetylcholine added to atropine | | |
| Vagus nerve stimulated while atropine + acetylcholine present | | |
| Control[1] | | |
| Epinephrine added | | |
| Control[1] | | |
| Heart removed from turtle | | |

[1] Previous treatments stopped or removed.
[2] Strength of contraction, part of the heart that beats, regularity of beat, etc.

# *Questions*

○ 1. Describe the effect of each of the treatments listed below on the heart:

a) stimulation of the vagus nerve _____

b) administration of acetylcholine _____

c) administration of atropine _____

d) administration of epinephrine _____

e) administration of atropine and acetylcholine

_____

f) stimulation of the vagus nerve plus administration of atropine and acetylcholine _____

○ 2. Where does contraction of the heart originate?

_____

○ 3. What inputs modify the frequency of depolarization of the heart muscle? _____

○ 4. What nerves are responsible for parasympathetic input to the heart? _____

5. What nerves are responsible for sympathetic input to the heart?

_____

○ 6. A child was accidentally poisoned by ingesting a pesticide containing an inhibitor of acetylcholinesterase. A physician administered atropine. Was this the correct antidote to use in this situation? Explain.

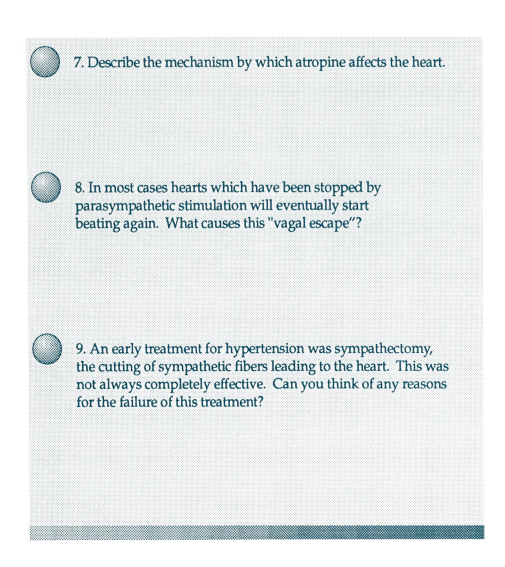

7. Describe the mechanism by which atropine affects the heart.

8. In most cases hearts which have been stopped by parasympathetic stimulation will eventually start beating again. What causes this "vagal escape"?

9. An early treatment for hypertension was sympathectomy, the cutting of sympathetic fibers leading to the heart. This was not always completely effective. Can you think of any reasons for the failure of this treatment?

# 16

# *Electrocardiogram*

*Reading assignment: text 306-309, 324-334*

## Objectives

 *Experimental*

1. To record electrocardiograms, the electrical signals produced by depolarization and repolarization of the muscles of the heart.

2. To determine the heart's electrical axis or cardiac vector.

 *Conceptual*

After completing this exercise and the reading assignment, you should be able to:

1. Explain how the electrical activity of the heart gives rise to electrical signals that are measurable at the surface of the body.

2. Explain how the events of the electrocardiogram are related to the corresponding mechanical events of the cardiac cycle.

3. Define the terms **diastole, P wave, QRS complex, atrioventricular node, AV delay, repolarization currents, T wave, cardiac vector, ectopic.**

4. Describe how the magnitude and direction of the cardiac vector are determined.

5. List the leads used in this exercise to record electrocardiograms and describe the placement of the electrodes of each lead.

6. Compare the arrangement of leads used in this exercise to the array of leads used for clinical purposes.

7. Describe and discuss the clinical significance of a) sinus arrhythmia and b) premature ventricular contraction.

8. Describe the method by which **Einthoven's triangle** is used to determine the direction and magnitude of the cardiac vector.

## Background

Extracellular fluid is an excellent conductor of electricity. When waves of depolarization and repolarization occur in heart muscle, small electrical currents flow between regions of the heart. These currents spread from the heart throughout the body. They are measurable as small differences in electrical potential between different spots on the body surface.

An **electrocardiogram** (ECG)(Fig. 16-1), made with a **polygraph** (also known as an **electrocardiograph**) records the sum of these electrical currents as registered by two active electrodes. The events of the electrocardiogram can be related to the events of the pump cycle of the heart (figs. 16-2, 16-3). The cardiac cycle is comprised of the following events:

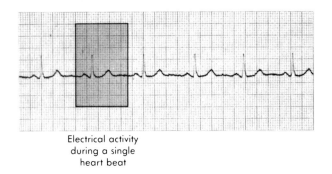

Electrical activity
during a single
heart beat

*Figure 16-1. A normal electrocardiogram.*

1. Diastole.  During **diastole**, the period when the ventricles are relaxed and both the atria and ventricles are filling with blood, there are no action potentials in any part of the heart.  No currents flow through the heart, and the pen of the electrocardiograph remains at its baseline position.  The ventricles are relaxed and are undergoing passive filling.

2. Depolarization of the atria.  As a wave of depolarization originating in the sinoatrial (SA) node spreads through the atria, current flows between the depolarized parts of the atria and adjacent parts that have not yet reached threshold.  These **depolarization currents** cause a deflection of the electrocardiogram pen called the **P wave**.

3. Atrial contraction.  This phase of the cardiac cycle begins when the atria are fully depolarized.  At this point the currents cease, the pen returns to baseline, and the P wave is over.  The resulting atrial contraction adds blood to the volume already in the ventricles.  The pen remains at baseline between the time the atria become fully depolarized and the time excitation spreads into the ventricles.

4. Atrioventricular delay.  A specialized node of conducting tissue, the **atrioventricular (AV) node**, provides a myocardial link between the atria and the ventricles (Fig. 16-4).  The ventricles do not become excited immediately after atrial contraction because of slowly conducting fibers immediately surrounding the AV node.  These fibers cause an **AV delay** of about 110 msec.

5. Ventricular depolarization.  The spread of depolarization into the ventricles is accompanied by depolarization currents that cause the second event of the ECG, the QRS complex.  This deflection is typically much larger than the P wave, because the mass of the ventricles is much greater than that of the atria.

When the entire muscle mass of the ventricles has entered the plateau phase of the cardiac action potential, current flow around the ventricles stops and the pen returns to baseline, ending the **QRS complex**.  Depolarization of the ventricles results in contraction that ejects some blood into the pulmonary arteries and aorta.

6. Ventricular repolarization.  Different parts of the ventricle begin to leave the plateau phase of their action potentials at slightly different times.  Currents flow between the repolarized regions and the still depolarized regions nearby.  These **repolarization currents** result in the third major event of the ECG, the **T wave**.  Similar, but much smaller, repolarization currents flow during atrial repolarization.  They are not normally seen in the ECG, however, because they are overwhelmed by the much larger currents of ventricular depolarization.

Repolarization is accompanied by loss of active tension in the ventricular muscle.  Ventricular pressure rapidly falls below arterial pressure, the pulmonary and aortic valves shut, the atrioventricular valves open, and the heart returns to diastole.

Placing the active electrodes at different positions on the body gives a two-dimensional picture of heart activity, because when the position of the active electrodes is changed, the electrodes "look" at the heart from a different angle.  Each electrode configuration is called a **lead**.

For clinical use electrocardiograms are recorded from a 12-lead array of electrode placement: the three bipolar leads that you will use, three **augmented leads** in which two of the limb leads are connected together, and six **precordial leads** in which an exploring electrode is attached to points

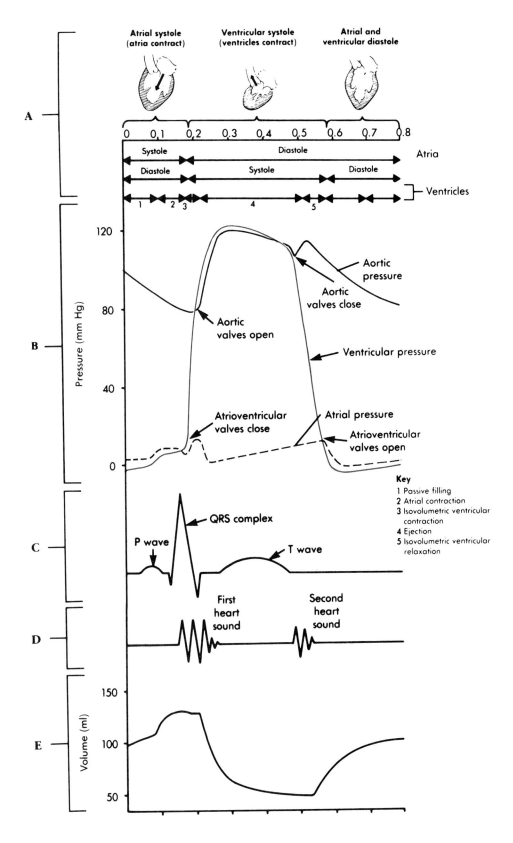

**A** — Atrial systole (atria contract) | Ventricular systole (ventricles contract) | Atrial and ventricular diastole

0   0,1   0,2   0,3   0,4   0,5   0,6   0,7   0,8

Systole | Diastole — Atria
Diastole | Systole | Diastole — Ventricles
1   2   3   4   5

**B**

Pressure (mm Hg)

120

80

40

0

Aortic pressure
Aortic valves close
Aortic valves open
Ventricular pressure
Atrioventricular valves close
Atrial pressure
Atrioventricular valves open

**Key**
1 Passive filling
2 Atrial contraction
3 Isovolumetric ventricular contraction
4 Ejection
5 Isovolumetric ventricular relaxation

**C**

QRS complex
P wave
T wave

**D**

First heart sound
Second heart sound

**E**

Volume (ml)

150

100

50

*Figure 16-2. The cardiac cycle. Phases and approximate durations are indicated at top of figure (A). The black, blue-green, and dashed curves show pressures in the aorta, left ventricle, and atrium, respectively (B). Below are shown the ECG (C) the heart sounds (D), and the ventricular volume (E). The phases of the cardiac cycle occur in the following order: (1) passive filling (early diastole), (2) atrial contraction (late diastole), (3) isovolumetric ventricular contraction, (4) ejection, and (5) isovolumetric ventricular relaxation.*

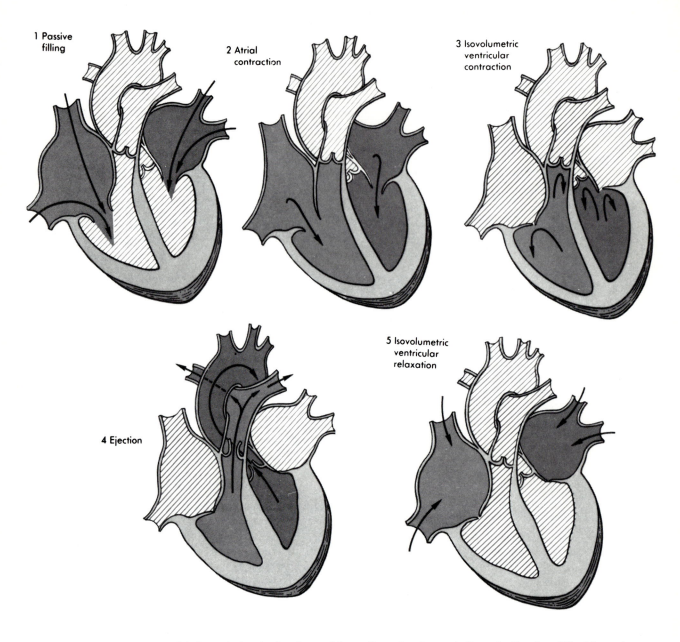

**Figure 16-3.** *Internal views of the heart during the five phases of the cardiac cycle. Arrows indicate the direction of blood flow.*

along a line that runs across the chest from the sternum to the left armpit. The 12-lead array gives more information of the type that you will obtain using a three-lead system.

In the system you will use, the voltage recorded at one limb is compared to the voltage measured at another limb, while the third limb serves as a ground. In **LEAD I** the active (positive) electrode is on the left arm, the passive (negative) electrode is on the right arm, and the ground is connected to the left leg (Fig. 16-5). With this configuration, upward deflection of the pen indicates that the left arm is electrically positive relative to the right arm. This is the condition that exists when a wave of depolarization or repolarization is moving toward the positive electrode.

In **LEAD II** the positive electrode is on the left leg, the negative electrode is on the right arm, and the

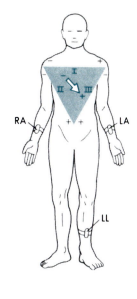

*Figure 16-5. The arrangement of the standard limb leads (I, II, and III). The axes of the three leads form Einthoven's triangle (shown in blue-green). The white arrow illustrates the mean electrical axis of the heart.*

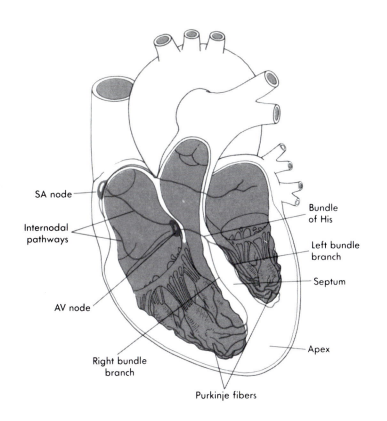

*Figure 16-4. The conducting fiber system of the heart. SA = sinoatrial; AV = atrioventricular.*

ground is connected to the left arm. In **LEAD III** the positive electrode is on the left leg, the negative electrode is on the left arm, and the ground is connected to the right arm.

A is a line having the properties of direction and magnitude. The **cardiac vector** represents the overall direction and magnitude of current flow around and within the heart at any instant. Any two of the three leads may be used to determine the cardiac vector.

In Part 1 of this exercise you will work with three partners to obtain ECGs for all members of your group. In Part 2 you will determine the cardiac vectors for your ECGs.

The durations of the different parts of the ECG contain important information. The mean inter-beat interval between the end of the T wave of one

cycle and the P wave of the next is a measure of heart rate. Ideally the interbeat interval should not vary from one beat to the next if the mean heart rate is not changing. If it does, an **arrhythmia** is present.

There are many types of arrhythmias, and the different types vary in significance. Two kinds of arrhythmia are occasionally seen in otherwise healthy people. A slight decrease of the interbeat interval with inspiration and increase with expiration is a **sinus arrhythmia**. It reflects changes in parasympathetic input to the heart during the respiratory cycle. Sinus arrhythmia is usually present in young adults, but its prevalence decreases with age.

**Premature ventricular contraction (PVC)** is another fairly common type of arrhythmia. These events can be caused by an **ectopic** ("out of place") ventricular pacemaker that occasionally triggers a ventricular contraction out of step with the atrial pacemakers. If this occurs during atrial diastole,

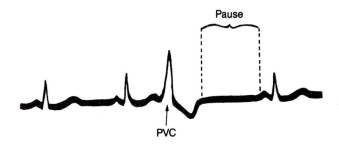

*Figure 16-6. A premature ventricular contraction (PVC).*

In a PVC the action potential does not follow the normal conduction pathway through the heart, so the corresponding ECG wave is broader and does not look like the QRS and T pattern of a normal ventricular contraction. Everyone probably experiences an occasional PVC, and occasional PVCs may be more common in athletes than in the general population. However, PVCs may be indicative of disease if they occur frequently (several per minute) or if they tend to occur during the T wave of normal cycles.

the ventricles may still be in their refractory period the next time a normal pacemaker signal reaches the ventricles. The ventricles will "skip a beat" (that is, the interbeat interval will be longer than usual) and undergo their next normal cycle after the next atrial contraction (Fig. 16-6).

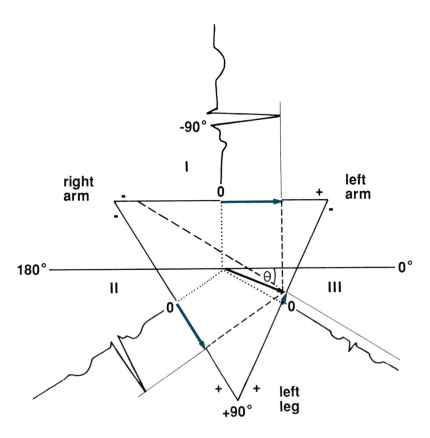

*Figure 16-7. Use of Einthoven's triangle to construct cardiac vector for the peak of the QRS complex.*

## *Procedure*

Record your results in the laboratory report section at the end of this exercise.

 *Part 1. Recording electrocardiograms*

1. The ECG will be most accurate if the subject lies quietly on a lab bench, because skeletal muscle activity introduces noise into the record. The subject should not be grounded outside of the electrodes. Do not touch water pipes, electrical conduit, or other students. Lie on a wooden bench if possible.

2. Set up the polygraph as demonstrated by your instructor. Place the leads as indicated in Fig. 16-5. Unless directed to do otherwise, set the paper speed at 25 mm/sec and the vertical scale at 1 mV/cm. These are the standard values for clinical recordings.

3. Each student should obtain a set of records from each of the three leads made at the same vertical gain. The record from each lead should contain at least several heart cycles.

4. If the pen goes off-scale at any point, decrease the amplifier gain and repeat previous measurements so that you obtain the necessary recordings.

5. While you are recording, look for a slight increase in the interval between beats that may occur with each breath.

 *Part 2. Determination of the cardiac vector*

1. Follow the procedure described in steps 2 to 5, below and shown in Fig. 16-7, to construct cardiac vectors for the peak of the QRS complex. Do this for each member of your group using the space provided in figs. 16-A to 16-D.

2. Using the traces obtained from at least two leads, measure the magnitude of the voltage deflection for the peak of the QRS complex.

3. Construct an equilateral triangle (Einthoven's triangle) representing the three leads. Draw perpendicular lines from the center of each side to determine the center of the triangle (Fig. 16-7: dotted lines).

4. For each lead, draw a line proportional to the measured deflection, from the center of the side of the triangle representing that lead toward the positive electrode of that lead if the deflection is upward, or toward the negative one if it is downward (Fig. 16-7: blue-green lines).

5. Next, draw a second pair of perpendicular lines (Fig. 16-7: dashed lines) from the ends of the magnitude lines toward the inside of the triangle. These will intersect inside the triangle. The cardiac vector (heavy black arrow) extends from the center of the triangle to the intersection of the magnitude lines.

Name: _____

Date: _____

Lab Section: _____

## *Analyzing Your Data*

Using the recording of your ECG, and those of your group members, identify the following: P wave, QRS complex, T wave, sinus arrhythmia (if present), premature ventricular contraction (if present).

Calculate heart rate, PR interval, QRS duration, QT interval, P amplitude, and T amplitude for each member of your group and enter these values in Table 16-1.

*Figure 16-A. Einthoven's triangle for member 1 of ECG group.*

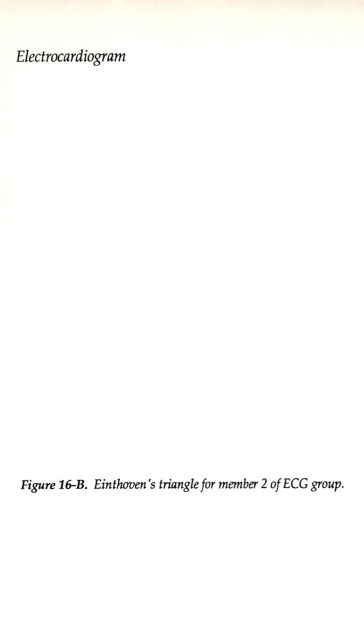

*Figure 16-B.* Einthoven's triangle for member 2 of ECG group.

*Figure 16-C.* Einthoven's triangle for member 3 of ECG group.

*Figure 16-D.  Einthoven's triangle for member 4 of ECG group.*

*Table 16-1.* Characteristics of electrocardiograms of group members.

| Characteristics of ECG | Group member | | | |
|---|---|---|---|---|
| | 1 | 2 | 3 | 4 |
| Heart rate | | | | |
| PR interval | | | | |
| QRS duration | | | | |
| QT interval | | | | |
| P amplitude | | | | |
| T amplitude | | | | |

# Questions

1. What causes the P wave of the ECG?

2. What causes the QRS complex of the ECG?

3. What causes the T wave of the ECG?

4. What is the atrioventricular node?

5. What causes AV delay?

6. What causes sinus arrhythmia?

7. What causes premature ventricular contractions?

8. What is the clinical significance of sinus arrhythmias and preventricular contractions?

9. What is a cardiac vector?

10. How is Einthoven's triangle used to determine the direction and magnitude of the cardiac vector?

11. Some people experience a condition called AV block, in which excitation fails to pass from the atria to the ventricles in some heart cycles. How would you recognize this condition in an ECG?

12. In some cases, a damaged branch bundle (Fig. 16-3) causes excitation to spread through some parts of the ventricles more slowly than normal. What effect would this have on the duration of the QRS complex?

13. The vectorcardiogram of the T wave usually has a similar orientation to that of the peak of the QRS complex. This means that the parts of the ventricle that are depolarized first are the (first, last) _____ to repolarize. Explain.

14. Freshly damaged heart muscle remains continuously depolarized. A current of injury flows between the damaged area and surrounding healthy muscle during diastole. What effect would this have on the baseline of the ECG during diastole?

15. If you were examining the trace from a single lead of the ECG for a freshly damaged heart, could you determine that a current of injury was flowing? _____ Explain.

# Human Respiration

*Reading assignment: text 408-412, 429-431, 451-457*

## Objectives

### Experimental

1. To determine your tidal volume, expiratory reserve volume, inspiratory reserve volume, and vital capacity using a spirometer.

2. To determine your tidal volume, expiratory reserve volume, inspiratory reserve volume, and vital capacity using a pneumograph.

3. To determine the effects on respiratory rate and on the ability to hold one's breath of a) hyperventilation and b) inhaling from a sack containing expired air.

### Conceptual

After completing this exercise and the reading assignment, you should be able to:

1. Define the terms **tidal volume, residual volume, inspiratory reserve volume, expiratory reserve volume, vital capacity, respiratory rate,** and **respiratory minute volume.**

2. Explain how measurements from a spirometer can be used to calibrate a pneumograph.

3. Discuss the clinical significance of vital capacity.

4. Describe the relationship between vital capacity and a) sex and b) body surface area.

5. Describe the effects of hyperventilation on respiratory rate and the ability to hold one's breath and explain how these effects are produced in terms of changes in arterial $P_{O_2}$ and $P_{CO_2}$ and blood pH.

6. Describe the effect of breathing from a sack containing expired air on respiratory rate and explain how this effect is produced in terms of changes in arterial $P_{O_2}$ and $P_{CO_2}$ and blood pH.

## Background

In earlier exercises you learned about the circulatory system that delivers oxygen to and removes carbon dioxide from the tissues. The link between the circulatory system and the atmosphere is the respiratory system. In this exercise you will measure pulmonary function, and you will also carry out some experiments on the regulation of breathing.

The changes in lung volume that occur during breathing can be measured with a **spirometer** (Fig. 17-1). In this method, the subject breathes from a tube connected to a floating inverted drum that rises as air enters it during expiration and falls as air leaves during inhalation.

A **capacity** is a measure of lung function that consists of two or more **volumes**. Total lung

A

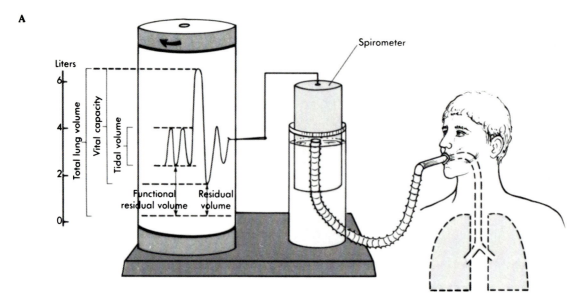

B

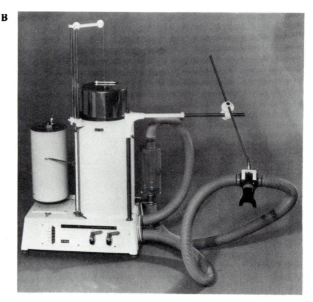

*Figure 17-1.  A Measurement of lung volumes by spirometry. Changes in lung volume are measured during normal quiet breathing and when a subject is asked to inhale maximally and then to exhale as much air as possible. B A spirometer.*

capacity is made up of the four volumes of air listed below (Fig. 17-2):

> \** **Tidal volume** is the amount of air that is inhaled and exhaled during a single breath.

> \** **Inspiratory reserve volume** is the amount of air that can be inhaled beyond the inhalation that occurs in quiet breathing.

> \** **Expiratory reserve volume** is the amount of air that can be forced out of the lungs after a normal, quiet exhalation would have ended.

> \** **Residual volume** is the air that remains in the lungs even after a forceful exhalation.

The total of tidal volume plus inspiratory and expiratory reserve volumes is known as **vital capacity**. This is the maximum amount of air that can be brought into the lungs and forcefully exhaled.

An individual's body size, sex, and physical condition affect vital capacity. In general, males of a given body size have a greater vital capacity than females of the same size. For women, vital capacity is approximately 2 l/m$^2$ of body surface area,

A

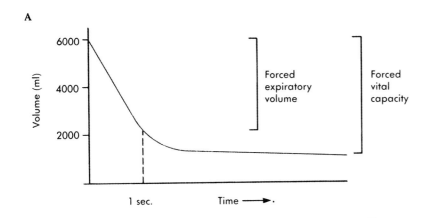

*Figure 17-2. Static lung volumes and capacities. A Tracing made by a spirometer. B Schematic view of lung.*

B

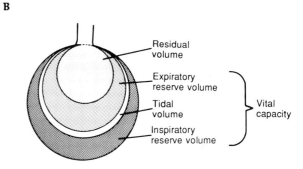

whereas for men it is approximately 2.5 l/m² of body surface area. (The estimation of vital capacity is more reliable if body surface area, rather than either height or weight, is used as an indicator of body size.)

Vital capacity is used as an index of the functional capacity of the lungs, although it is no longer as popular in clinical diagnosis as it once was. In certain disease states vital capacity is altered; for example, pneumonia or emphysema may be accompanied by a decrease in vital capacity. Conditions that decrease the lung space in the thoracic cavity, such as the presence of a tumor or the enlargement of blood vessels due to heart disease, can also decrease vital capacity.

You will measure tidal volume and vital capacity directly. Expiratory and inspiratory reserve volumes cannot be measured directly but can be calculated as follows:

Expiratory reserve volume = (Volume of air expired in a forceful exhalation after a normal breath) – (Tidal volume).

Inspiratory reserve volume = (Vital capacity) – (Tidal volume) – (Expiratory reserve volume).

The number of breaths taken per unit time is the **respiratory rate**. The product of respiratory rate in breaths/min and tidal volume is the volume of air entering and leaving the respiratory system in one minute, or the **respiratory minute volume**.

In Part 1 of this exercise you will measure pulmonary volumes with a **spirometer**. In Part 2 a chart recording of respiratory movements will be made using a **pneumograph** (Fig. 17-3). Using a known value for vital capacity determined with the spirometer, you will be able to calibrate pneumograph recordings of your breathing patterns. You will use this information in Exercise 18.

Breathing is a rhythmic process controlled by several factors. The **partial pressure** of a gas is that part of the total pressure of a gas mixture which is attributable to that gas. Dry air is about 21% $O_2$. Therefore, the partial pressure of oxygen ($P_{O_2}$) in dry air at sea level is:

$$760 \text{ mm Hg} \times 0.21 = 160 \text{ mm Hg}.$$

At normal or nearly normal $P_{O_2}$, changes in the arterial $P_{CO_2}$ have a much greater effect on respiratory minute volume than do variations in the arterial $P_{O_2}$. The respiratory response to $CO_2$ originates in the brain. Respiratory chemoreceptors in the brain do not detect the plasma $P_{CO_2}$ directly.

**Figure 17-3.** *A pneumograph.*

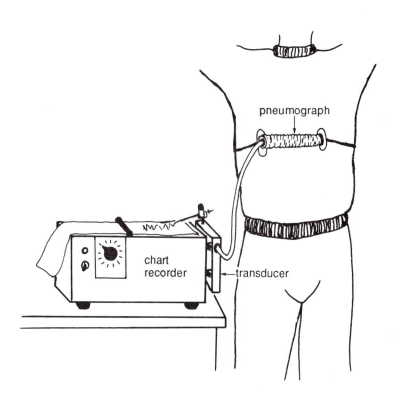

Instead, they respond to the pH of the **cerebrospinal fluid** (CSF). The walls of the capillaries in the brain constitute a **blood-brain barrier** that is not freely permeable to ions such as $H^+$ or $HCO_3^-$. However, dissolved $CO_2$ crosses the barrier readily. Any increase in the arterial $P_{CO_2}$ results in an immediate and equivalent increase in the $P_{CO_2}$ of the CSF.

Once in the CSF, the $CO_2$ reacts with water to form carbonic acid ($H_2CO_3$), which dissociates to produce $H^+$ and $HCO_3^-$ (Fig. 17-4). Compared with the blood, CSF is weakly buffered. The low buffer capacity of the CSF makes it sensitive to changes in $P_{CO_2}$. When the arterial $P_{CO_2}$ increases, the $P_{CO_2}$ of the CSF increases proportionately, and the subsequent reaction of $CO_2$ and water decreases CSF pH. The response of the central chemoreceptors to this change in pH increases the respiratory drive.

Although the arterial $P_{CO_2}$, as monitored by the central chemoreceptors, is responsible for most of the respiratory drive, the respiratory centers are also responsive to changes in the arterial $P_{O_2}$ and arterial pH detected by **peripheral chemoreceptors**. The major peripheral chemoreceptors are located in the carotid and aortic bodies (Fig. 17-5).

In Part 3 you will investigate the effects of breath-holding and hyperventilation on respiratory rate. When you hold your breath, plasma $P_{O_2}$ falls and $P_{CO_2}$ rises. However, when breathing is preceded by a period of rapid, deep breathing (**hyperventilation**), arterial $P_{CO_2}$ is lower than normal at the start of breath-holding (Fig. 17-6), and the period of breath-holding increases. Before the $P_{CO_2}$ rises enough to trigger inspiration, the arterial $P_{O_2}$ may fall so low that a decrease in the $O_2$ supply to the brain triggers unconsciousness. There are a number of reported cases in which swimmers who hyperventilated before swimming underwater have become unconscious and drowned.

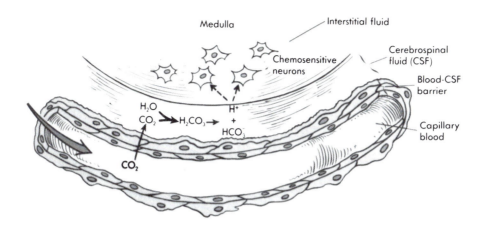

*Figure 17-4. Activation of central chemoreceptors by changes in the arterial* $P_{CO_2}$. *Although the receptors are sensitive to small changes in the* $P_{CO_2}$, *they are not directly stimulated by* $CO_2$. *Instead they respond to changes in the pH of the cerebrospinal fluid.* $CO_2$ *diffuses across the blood-brain barrier and forms carbonic acid, which dissociates. By this mechanism, an increase in the arterial* $P_{CO_2}$ *acidifies the CSF.*

*Figure 17-5. The locations of carotid and aortic arch chemoreceptors.*

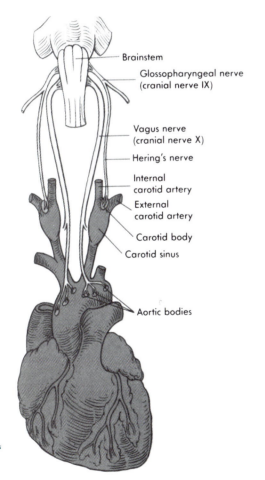

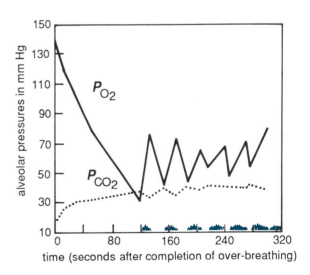

*Figure 17-6. Changes in* $P_{CO_2}$ *and* $P_{O_2}$ *following hyperventilation. Blue-green lines indicate breathing movements. (Adapted from Douglas and Haldane.)*

## *Procedure*

Record your results in the laboratory report section at the end of this exercise.

 *Part 1. Respiratory volumes*

1. Swab the mouthpiece of the spirometer with 80% ethanol.

2. Practice exhaling into the instrument while wearing nose clips. Do not begin recording until you have become accustomed to the nose clips.

3. While standing up, place the indicator dial at the zero mark. Take in a normal breath and exhale normally through the mouthpiece. This is a measurement of your tidal volume. Do <u>not</u> watch the dial of the instrument while making the measurement.

4. Repeat Step 3 five more times. Enter your readings below. Compute your average tidal volume and enter this value in Table 17-1A.

5. While standing up, take a normal breath and then exhale as completely as you can into the spirometer. Enter this value for maximum exhalation in Table 17-1B.

6. Repeat Step 5 two more times or until consistent consecutive measurements are obtained. Enter these values in Table 17-1B and compute the average of these readings.

7. To determine expiratory reserve volume, subtract mean tidal volume from the average value for maximal exhalation. Enter this value in Table 17-1C.

8. To determine vital capacity, take in as much air as possible by a forced inspiration. Expire through the mouth piece, emptying the lungs as completely as possible. Enter this value in Table 17-1D.

9. Repeat Step 8 two more times. Compute the mean of the three readings for vital capacity and enter this value in Table 17-1D..

10. Repeat steps 8 and 9 in a sitting position. Compute mean vital capacity in a sitting position and enter this value in Table 17-1E.

11. Repeat steps 8 and 9 in a reclining position. Compute mean vital capacity in a reclining position and enter this value in Table 17-1F.

12. Calculate your inspiratory reserve volume. Enter this value in Table 17-1G.

In parts 2 and 3 you should work with a partner. Take turns being the subject and the experimenter.

 *Part 2. Recording respiratory movements*

1. Set up a pneumograph recording module apparatus and the event/time marker module (Fig. 17-3). The pneumograph should be positioned at about mid- to low-chest level. Do not stretch the pneumograph too much. Note that pressure in the system can be regulated by letting air in and out with the relief valve on the module. The system should be open to the atmosphere. Do not close the relief valve while positioning the pneumograph on the body; if the system is closed, the neoprene diaphragm in the module is likely to break.

2. Subject: stand and face away from the recorder.

3. Experimenter: record a few normal inhalations and exhalations and, while doing so, adjust the speed of the chart paper so that the peaks of the waves are about 5 mm apart. Write what is being recorded on the chart paper in each series of recordings.

4. Record a maximal inhalation followed by a maximal exhalation. The tracing produced will allow you to measure vital capacity, but the pneumograph does not allow you to read the volume directly. To do this, you will need to calibrate your pneumograph reading with the readings previously obtained with the spirometer. This recording and the earlier reading obtained from the spirometer will allow you to convert your pneumograph records to volumes of air.

5. To calculate the conversion factor, divide the vital capacity reading obtained with the spirometer (ml) by the height of the pneumograph recording of vital capacity (mm). The answer indicates the number of ml of air represented by a height of 1 mm on the recording. This is your conversion factor, in ml/mm. It tells you how many ml of air are represented by 1 mm of pneumograph tracing. Use it to convert distance on the pneumograph recording (in mm) to volumes (in ml). For example, if you measured a vital capacity of 4,000 ml using the spirometer, and your recording for vital capacity on the chart paper was 40 mm, your conversion factor would be:

4,000 ml/40 mm = 100 ml/mm.

In this example, a volume of 100 ml would be indicated by a vertical distance of 1 mm on the pneumograph recording. Record the conversion factor in Table 17-2.

6. On the chart paper, label portions of the vital capacity recording (in mm) that consist of inspiratory reserve volume, tidal volume, and expiratory reserve volume (Fig. 17-2A). Enter these volumes in Table 17-2.

7. Convert your pneumograph readings for each of these fractions to volumes and enter this information in Table 17-2. Compare these values to the values you obtained with the spirometer (Table 17-1).

 *Part 3. Regulation of breathing pattern*

1. Subject: make sure the relief valve of the pneumograph is open initially. Put the pneumograph on and sit down. Do not watch the chart during any of the experiments.

2. Experimenter: set the chart speed so that the peaks and valleys recorded as the subject breathes are clearly separated from each other. Note the chart speed.

3. Subject: breathe normally for 1 min, then expire and hold your breath for as long as possible.

4. Experimenter: record the subject's respiratory movements for 30 sec after breathing begins again.

5. Subject: take about ten deep breaths, expiring rapidly each time, then expire and hold your breath for as long as possible.

6. Experimenter: open the side arm of the pneumograph tubing and stop the recording during the subject's hyperventilation but reconnect and start the recorder instantly at the end of hyperventilation. The recording will then show the respiratory pattern after hyperventilation. Continue recording until rate and depth of breathing have returned to normal.

7. Subject: breathe normally for 15 sec, then breathe into a paper sack for 2 min. Be sure that all air comes from and returns to the sack.

8. Experimenter: record the subject's respiration until there is a noticeable change in the breathing pattern.

# Laboratory Report

### Exercise 17: Human Respiration

Name: _____

Date: _____

Lab Section: _____

## Analyzing Your Data

Your chart recordings and the data in the tables summarize your data for this exercise. Be sure you understand how the different respiratory volumes are related to each other and how the pneumograph was calibrated. You should be able to interpret your results from Part 3 in terms of what is happening to arterial $P_{O_2}$ and $P_{CO_2}$.

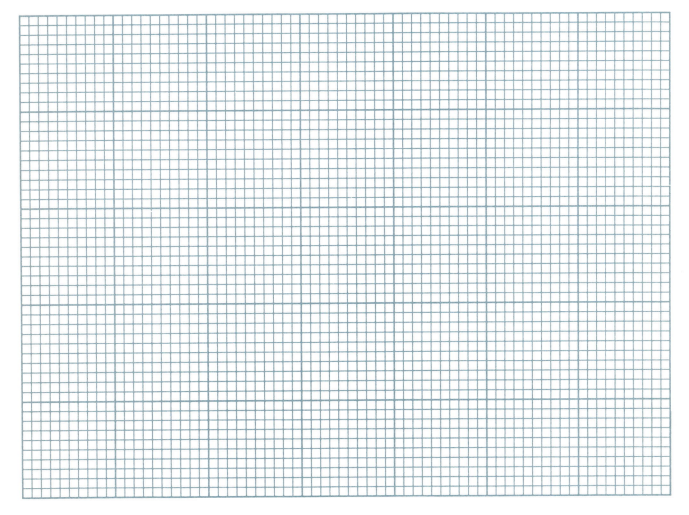

**Table 17-1.** *Respiratory volumes and vital capacity determined from spirometry.*

### A. Readings for tidal volume (ml).

Average:

### B. Readings for maximal exhalation (ml).

Average:

### C. Calculation of expiratory reserve volume (ml).

| Average maximal exhalation: | — | Mean tidal volume: | = | Expiratory reserve volume: |

**Table 17-1**, *continued*

**D. Readings for vital capacity (standing).**

Average:

**E. Readings for vital capacity (sitting).**

Average:

**F. Readings for vital capacity (reclining).**

Average:

**G. Calculation of inspiratory reserve volume.**

| Average vital capacity (standing): | Expiratory reserve volume: | Tidal volume: | Inspiratory reserve volume: |
|---|---|---|---|
| | − | − | = |

**Table 17-2.** *Respiratory volumes measured with a spirometer and a pneumograph.*

Vital capacity measured with spirometer:        ml

Height of pneumograph tracing representing vital capacity:        mm

Conversion factor:        ml/mm.

| Respiratory volume | Height of pneumograph recording (mm) | Volume of air estimated with pneumograph (ml)[1] |
|---|---|---|
| Tidal volume | | |
| Expiratory reserve volume | | |
| Inspiratory reserve volume | | |

[1] Height of pneumograph recording (mm) x conversion factor (ml/mm).

*Table 17-3. Regulation of breathing.*

## A. Regulation of inspiration.

| Condition preceding holding of breath | Time breath held (sec) |
|---|---|
| Normal breathing | |
| Hyperventilation | |

## B. Regulation of respiratory rate.

| Condition preceding measurement of respiratory rate | Respiratory rate (breaths/min) |
|---|---|
| Breathing normally | |
| Holding breath for as long as possible (see A, above) | |
| Hyperventilating | |
| Breathing into and from a paper sack | |

## Questions

1. What is the difference between a respiratory volume and a capacity?

2. Fig. 17-A is a tracing from a spirometer. What do a, b, c, d, and e represent?

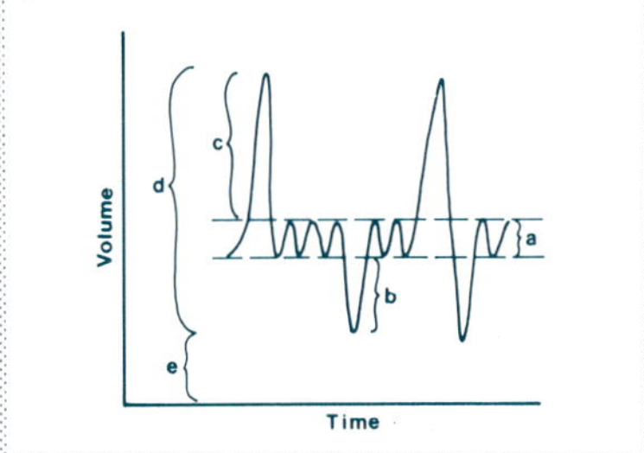

*Figure 17-A.*

a) _____    b) _____

c) _____    d) _____

e) _____

3. What is the clinical significance of vital capacity?

4. What is meant by the "partial pressure" of a gas?

5. Chemoreceptors in the brain respond to changes in the pH of the cerebrospinal fluid, but the blood-brain barrier blocks the passage of most ions, including $H^+$, into the CSF. What, then, causes the changes in pH to which the cerebral chemoreceptors respond?

6. Where are the peripheral chemoreceptors located? What stimulates them?

7. Was the time you were able to hold your breath after you hyperventilated longer than when you had not been hyperventilating? Explain why, in terms of the effects of hyperventilation on arterial $P_{O_2}$, $P_{CO_2}$, and blood pH.

8. How did expiring into and inspiring from a paper sack affect your respiratory rate? Describe the mechanism that produces this result.

# Human Heart Rate and Blood Pressure

*Reading assignment: text 334-343, 362-363, 376-381, 386-392, 460-461*

## Objectives

 *Experimental*

1. To determine the effects of posture on blood pressure and pulse rate.

2. To measure respiratory and cardiovascular responses to mild and strenuous exercise.

 *Conceptual*

After completing this exercise and the reading assignment, you should be able to:

1. Explain how blood pressure is measured with a sphygmomanometer and a stethoscope.

2. Describe the effects of posture on blood pressure and pulse rate and the mechanisms that produce these effects.

3. Define the terms **cardiac output, arterial systolic pressure, systole, Korotkoff sounds, arterial diastolic pressure, pulse pressure, stroke volume,** and **feed-forward mechanism.**

4. Describe the location of the **baroreceptors** and how the reflex response of these receptors regulates blood pressure.

5. Describe the cardiovascular responses to exercise and the mechanisms that produce these responses.

6. Describe the mechanisms that are thought to produce the respiratory response to a) moderate and b) strenuous exercise.

## Background

Gas exchange between blood and tissue depends on pulmonary ventilation and blood circulation. In this exercise you will examine cardiovascular responses to changes in posture (Part 1) and you will investigate the parallel responses of the respiratory and circulatory systems to the increased $O_2$ demands of exercise (Part 2).

**Cardiac output,** the volume of blood pumped by a single ventricle per minute, is an indicator of how much work the heart is doing. It is not practical to measure cardiac output directly. Instead you

will use two indirect measures of the work of the heart: blood pressure and heart rate. There are several methods that can be utilized to measure blood pressure. Physicians commonly use a **sphygmomanometer** and a **stethoscope.** The sphygmomanometer is a device for measuring pressure, consisting of a calibrated mercury manometer and a hollow cuff equipped with a pressure bulb (Fig. 18-1). The cuff is wrapped around the upper arm and the pressure bulb is used to inflate the cuff to a pressure greater than **arterial systolic pressure,** the maximum arterial pressure during ventricular contraction (**systole**). The pressure level is read from a mercury

*Figure 18-1. Measurement of blood pressure with a sphygmomanometer.*

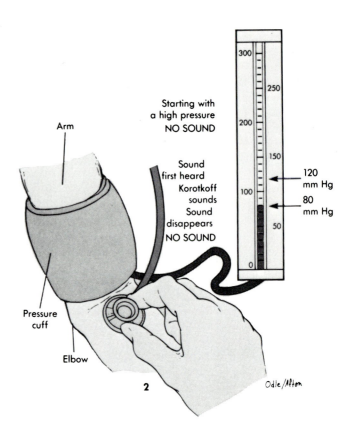

manometer. The high pressure in the cuff is transmitted through the arm and causes the **brachial artery** to collapse at around 200 mm Hg, preventing blood from flowing through the artery.

After the cuff has been inflated to 200 mm Hg, the stethoscope is placed over the brachial artery just proximal to the elbow joint. The air in the cuff is then slowly released by loosening the screw on the pressure bulb. This causes the pressure in the cuff and the arm to drop. The stethoscope is used to detect the first sound of the blood being forced through the partially occluded artery. When the pressure in the cuff falls slightly below the systolic pressure, arterial pressure at the peak of systole is greater than the pressure in the cuff. At this time blood can spurt through the artery only when arterial pressure is at its highest value. The flow of blood through the partially occluded artery occurs at a very high velocity, producing vibration and turbulence that can be heard with the stethoscope as soft tapping sounds, called **Korotkoff**

**sounds**. The pressure reading on the manometer when these sounds are first heard corresponds to arterial systolic pressure and indicates the pressure during ventricular systole. Arterial systolic pressure is typically around 120 mm Hg.

As the pressure in the cuff continues to decrease, the tapping sounds first become louder and more distinct, as more blood flows through the artery. Then the sounds become muffled. At this point the artery is open throughout the cycle, and the flow of blood is continuous but still turbulent. When the cuff pressure falls to arterial diastolic pressure, typically around 70 mm Hg, blood flow through the artery becomes nonturbulent, and the sounds disappear. The reading at this point is the **arterial diastolic pressure**. It indicates the pressure in the heart and large arteries during ventricular diastole. The difference between the systolic pressure and the diastolic pressure is called the **pulse pressure**.

Blood pressure, which is influenced by both the pumping action of the heart and the degree of constriction or dilation of blood vessels, is under reflex control. When you are lying down, the mean arterial blood pressure of the major arteries in your head, trunk, and legs is similar; the same is true for the pressure in your large veins (Fig. 18-2). When you stand up, gravity causes arterial pressure to decrease in the upper parts and increase in the lower parts of your body. Standing up from a prone position has an effect on blood pressure that is equivalent to losing approximately 500 ml of blood. In the absence of reflex responses to this situation, the decrease in venous return of blood to the heart would soon result in a decrease in cardiac output and fainting would result.

Standing up without fainting is possible because of the **baroreceptor reflex**. When arterial blood pressure drops, stretching of the arterial walls decreases, resulting in a decrease in the rate at which **baroreceptors** in the **carotid sinus** and the **aortic arch** (Fig. 18-3) discharge action potentials. The medullary cardiovascular control center responds by increasing sympathetic and decreasing parasympathetic inputs to the heart and blood vessels (Fig. 18-4). As a consequence of these changes, the heart increases its rate and force of contraction, arterioles constrict, and arterial blood pressure is restored. In addition, contraction of muscles of the limbs and the respiratory system aids venous return of blood to the heart.

To maximize blood flow to muscles during exercise, cardiac output must be increased, and the working muscles must accomplish the largest possible share of the increased output. Cardiac output can be increased by increasing **stroke volume** (the amount of blood ejected from a ventricle in one heartbeat), **heart rate**, or both.

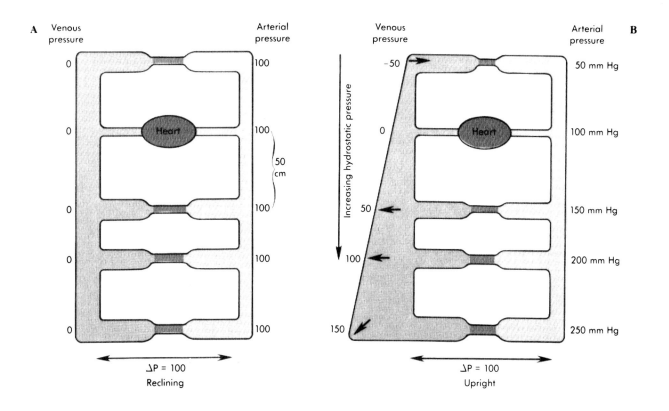

*Figure 18-2. Pressures in the cardiovascular system for a person who is **A**, lying down and **B**, standing. The pressures in mm Hg are given at various levels in both the arteries and the veins.*

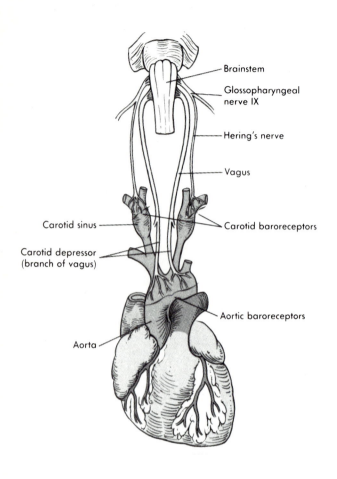

*Figure 18-3. The location of carotid and aortic barorecep-tors and their pathways into the central nervous system.*

During exercise, increases in heart rate and stroke volume result in increased cardiac output (Fig. 18-5). In addition, dilation of arterioles serving the skeletal muscles results in a large increase in the share of the cardiac output that is delivered to the muscles (Fig. 18-6). Because of this dilation of muscle arterioles, total peripheral resistance usually falls during moderate-to-heavy exercise. This causes an increase in venous return and, as a result, a further increase in cardiac output.

During moderate exercise, systolic pressure usually increases (Fig. 18-7). At the same time, the decrease in total peripheral resistance speeds blood runoff from the aorta into the arterial tree. This allows aortic pressure to fall rapidly between beats. Diastolic pressure may even decrease with

increasing workload in well-muscled people. The net effect of these changes is to leave mean arterial pressure virtually unchanged. Thus, changes in the driving force for blood flow are not important in increasing cardiovascular performance during exercise.

The mechanisms that trigger the respiratory response to exercise are presently not well understood. The dramatic increase in respiratory minute volume that occurs during exercise is not triggered by changes in arterial $P_{O_2}$, $P_{CO_2}$, or plasma pH. In fact, arterial $P_{O_2}$, $P_{CO_2}$, and plasma pH do not change very much during moderate exercise (Fig. 18-8). It has been suggested that **feedforward** regulation, in which the body's need for increased gas exchange is anticipated, may be involved in the increase in ventilation at the start of exercise.

However, plasma pH does play a role in the regulation of the respiratory response to strenuous exercise. When the demand for $O_2$ exceeds the body's capacity for $O_2$ delivery, an oxygen debt develops. At this point, stimulation of the peripheral chemoreceptors by increasing levels of plasma lactic acid causes a substantial increase in ventilation. This additional stimulation of ventilation occurs in spite of the decrease in arterial $P_{CO_2}$ and the increase in arterial $P_{O_2}$ caused by the initial increase in ventilation (Fig. 18-8).

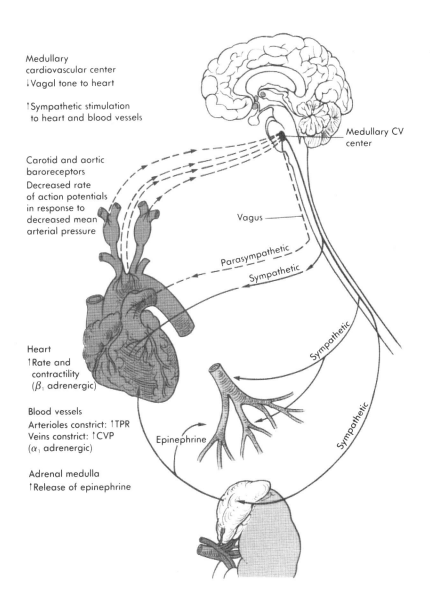

Medullary
cardiovascular center
↓Vagal tone to heart

↑Sympathetic stimulation
to heart and blood vessels

Medullary CV
center

Carotid and aortic
baroreceptors
Decreased rate
of action potentials
in response to
decreased mean
arterial pressure

Vagus

Parasympathetic

Sympathetic

Sympathetic

Heart
↑Rate and
contractility
($\beta_1$ adrenergic)

Blood vessels
Arterioles constrict: ↑TPR
Veins constrict: ↑CVP
($\alpha_1$ adrenergic)

Epinephrine

Sympathetic

Adrenal medulla
↑Release of epinephrine

*Figure 18-4.* *Baroreceptor response to a decrease in mean arterial pressure.* CV = *cardiovascular,* TPR = *total peripheral resistance,* CVP = *central venous pressure.*

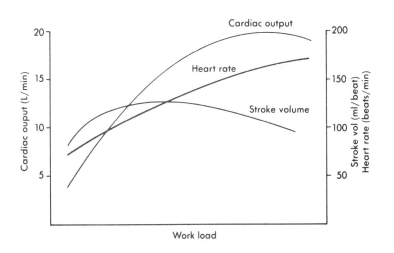

*Figure 18-5.* *Changes in heart rate and stroke volume (right vertical scale) and their product, cardiac output (left vertical scale) as workload (horizontal axis) is increased during exercise. Note that heart rate increases continuously as the workload is increased, but stroke volume first becomes stable and then decreases as the heart rate becomes very high. At the highest workloads, cardiac output is limited by the decreasing stroke volume.*

273

*Figure 18-6. Typical changes in organ perfusion with exercise. The left bar in each pair shows the resting blood flow; the right bar shows the flow during exercise.*

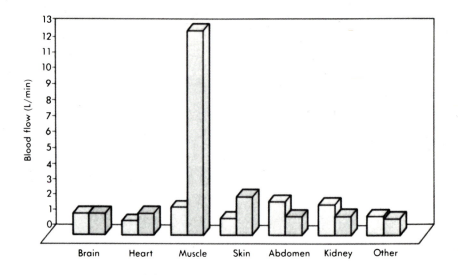

*Figure 18-7. The effect of increasing workload on systolic pressure, diastolic pressure, and mean arterial pressure.*

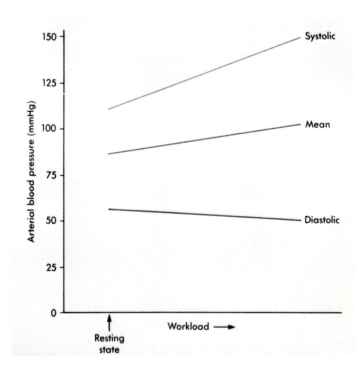

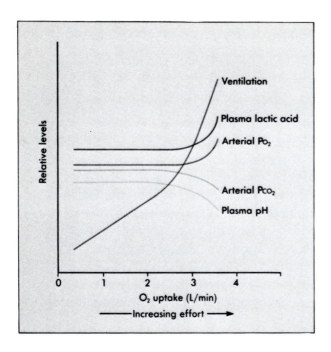

*Figure 18-8. Changes in blood gas composition, plasma pH, and plasma lactic acid levels with increasing effort, as measured by the rate of O$_2$ uptake. The beginning of oxygen debt is indicated by the inflections of the curves.*

## Procedure

Record your results in the laboratory report section at the end of this exercise.

Work in pairs. Take turns being subject and experimenter.

 **Part 1. Cardiovascular responses to changes in posture**

1. Before beginning, both partners should practice taking pulse rate and blood pressure readings, so that they can be taken rapidly and accurately.

Recall that systolic pressure is taken at the time sounds are first heard, and diastolic pressure is taken when the sounds first disappear. <u>The cuff of the sphygmomanometer should not be left on the arm for extended periods of time</u>, because it impedes the flow of blood to the forearm.

The pulse can be felt at several points on the surface of the body. The most convenient point is on the underside of the wrist. Use your <u>index finger</u> to detect the pulse of the subject and practice a few times so that you can locate it reliably. To take your partner's pulse, count the number of beats in a 6 sec period and multiply this value by 10. This will give you the number of beats in one minute.

2. Subject: lie on your back on a lab table.

3. Experimenter: after the subject has been lying quietly for five minutes, take the subject's blood pressure and pulse. Enter these data in Table 18-1.

4. Subject: stand up.

5. Experimenter: as rapidly as possible, measure the subject's pulse rate and then take the subject's blood pressure. (You have to be fast here to detect the changes.) Enter these values in Table 18-1.

6. Repeat Step 5 after the subject has been standing for five minutes. Continue repeating this procedure at 2 min intervals until pulse rate and blood pressure are stable (unchanging). Enter your results in Table 18-1.

 *Part 2. Cardiovascular and respiratory responses to exercise*

Only one partner should exercise on the exercycle.

1. Subject: classify yourself as in "good physical condition" if you participate in sports or exercise daily or in "average or below-average condition." (Be honest.) Sit on the exercycle, wearing the pneumograph, strapped below your rib cage, and the blood pressure cuff.

2. Experimenter: record 2 min of resting ventilation. Use a chart speed of 0.05 cm/sec for all measurements of respiration rate and depth before and after exercising. During exercise, increase the chart speed to 0.1 cm/sec. Write the chart speed you used at appropriate places on the recording. At the end of the 2 min period take resting blood pressure and pulse rate. Enter these results in Table 18-2.

3. Subject: adjust the speed and resistance of the exercycle. Speed should be set at 60 rpm. Resistance will be measured either in kilogram-meters per minute (kg-m/min) or watts. (1 watt = 6.1 kg-m/min.)

4. Subject: begin by pedaling with the resistance set at 600 kg-m/min (100 watts) for 2 min. This is your period of moderate exercise. Without stopping, increase the resistance to 900 kg-m/min (150 watts) and pedal at the same rate for another 2 min. This is your period of strenuous exercise.

5. Experimenter: record the subject's respiration throughout both periods of exercise.

6. Experimenter: as soon as possible after the subject stops exercising, measure pulse rate and then blood pressure.

7. Experimenter: repeat the pulse rate measurement every 2 min and blood pressure determination every 3 min until they return to normal. Enter these values in Table 18-2. In addition, continue recording the subject's ventilation pattern during the recovery period.

8. When the subject has recovered from exercising, use the chart recording to determine respiratory minute volumes (see Exercise 17) for each 2-min interval beginning with the 2-min rest period before the onset of exercise. To do this you will need the conversion factor you determined in Exercise 17. This will tell you how many ml of air are represented by each mm on your pneumograph tracing. Since respiratory minute volume is the volume of air breathed per minute, you will need to divide the volume of air taken in during each of the 2-min intervals by 2 to obtain respiratory minute volume. Enter these values in Table 18-3.

9. Enter your data on the board. Using pooled data from the class, calculate average values for blood pressure, heart rate, and respiratory minute volume for a) people who are in good physical condition and b) people who are in average or below-average physical condition. Enter these values in tables 18-4 and 18-5.

# Laboratory Report

### *Exercise 18: Human Heart Rate and Blood Pressure*

Name: _____

Date: _____

Lab Section: _____

## *Analyzing Your Data*

Using the pooled class data from Table 18-4, make a graph showing the relationship between heart rate (independent variable) and post-exercise time (dependent variable) for conditioned and unconditioned subjects. In addition, make a second graph showing changes in systolic and diastolic pressure as a function of post-exercise time in conditioned and unconditioned subjects. This graph will have four lines; be sure they are all clearly labeled. Finally, graph respiratory minute volume for conditioned and unconditioned subjects, using the data in Table 18-5.

**Table 18-1.** *The effect of posture on blood pressure.*

| Posture | Blood pressure (mm Hg)[1] | Pulse rate (beats/sec) |
|---|---|---|
| Prone | | |
| Standing (10 sec after rising) | | |
| Standing (5 min after rising) | | |
| Standing (7 min after rising) | | |
| Standing (9 min after rising) | | |

[1] Systolic pressure/diastolic pressure.

277

Table 18-2. *The effect of exercise on blood pressure and heart rate.*

| Experimental conditions | Blood pressure (mm Hg)[1,2] | Heart rate[3] |
|---|---|---|
| 2 min at rest | | |
| Immediately after 2 min moderate exercise and 2 min heavy exercise | | |
| 2 min after heavy exercise | – | |
| 3 min after heavy exercise | | – |
| 4 min after heavy exercise | – | |
| 6 min after heavy exercise | | |
| 8 min after heavy exercise | – | |
| 9 min after heavy exercise | | – |

[1] Taken at end of 3 min period.
[2] Systolic pressure/diastolic pressure.
[3] Taken at the end of 2 min period.

*Table 18-3. The effect of exercise on respiratory minute volume.*

| Time (min) from start of experiment | Activity | Respiratory minute volume (liters/min) | Respiration rate (breaths/min) |
|---|---|---|---|
| 0-2 | At rest | | |
| 2-4 | Moderate exercise | | |
| 4-6 | Heavy exercise | | |
| 6-8 | Recovery | | |
| 8-10 | Recovery | | |
| 10-12 | Recovery | | |

*Table 18-4. Comparison of the effect of exercise on blood pressure and heart rate in conditioned and unconditioned subjects.*

| Experimental conditions | Blood pressure (mm Hg)[1,2] | | Heart rate[3] | |
|---|---|---|---|---|
| | Con. | Ucon.[4] | Con. | Ucon.[4] |
| 2 min at rest | | | | |
| Immediately after 2 min moderate exercise and 2 min heavy exercise | | | | |
| 2 min after heavy exercise | – | – | | |
| 3 min after heavy exercise | | | – | – |
| 4 min after heavy exercise | – | – | | |
| 6 min after heavy exercise | | | | |
| 8 min after heavy exercise | – | – | | |
| 9 min after heavy exercise | | | – | – |

[1] Taken at end of 3 min period.
[2] Systolic pressure/diastolic pressure.
[3] Taken at end of 2 min period..
[4] Con. = conditioned; Ucon. = unconditioned.

*Table 18-5. Comparison of the effect of exercise on respiratory minute volume in conditioned and unconditioned subjects.*

| Time (min) from start of experiment | Activity | Respiratory minute volume (liters/min) | | Respiration rate (breaths/min) | |
|---|---|---|---|---|---|
| | | Con. | Ucon.[1] | Con. | Ucon.[1] |
| 0-2 | At rest | | | | |
| 2-4 | Moderate exercise | | | | |
| 4-6 | Heavy exercise | | | | |
| 6-8 | Recovery | | | | |
| 8-10 | Recovery | | | | |
| 10-12 | Recovery | | | | |

[1] Con. = conditioned; Ucon. = unconditioned.

## *Questions*

1. What changes in heart rate and blood pressure did you observe when you stood up after a period of lying down? _____ What reflex pathway caused these changes?

2. What are baroreceptors? Where are they located?

3. List two ways that cardiac output can be increased. _____

4. What role do baroreceptors play in the cardiovascular response to exercise?

5. How is the blood supply to skeletal muscles increased during exercise?

6. A trained athlete will show a much less dramatic change in circulatory and respiratory function in response to exercise than someone who has not undergone training. Why?

7. The mechanism that produced changes in ventilation in response to moderate exercise is not well understood. Describe a mechanism that has been suggested for this response.

8. How is the respiratory response to strenuous exercise regulated?

Human Heart Rate and Blood Pressure

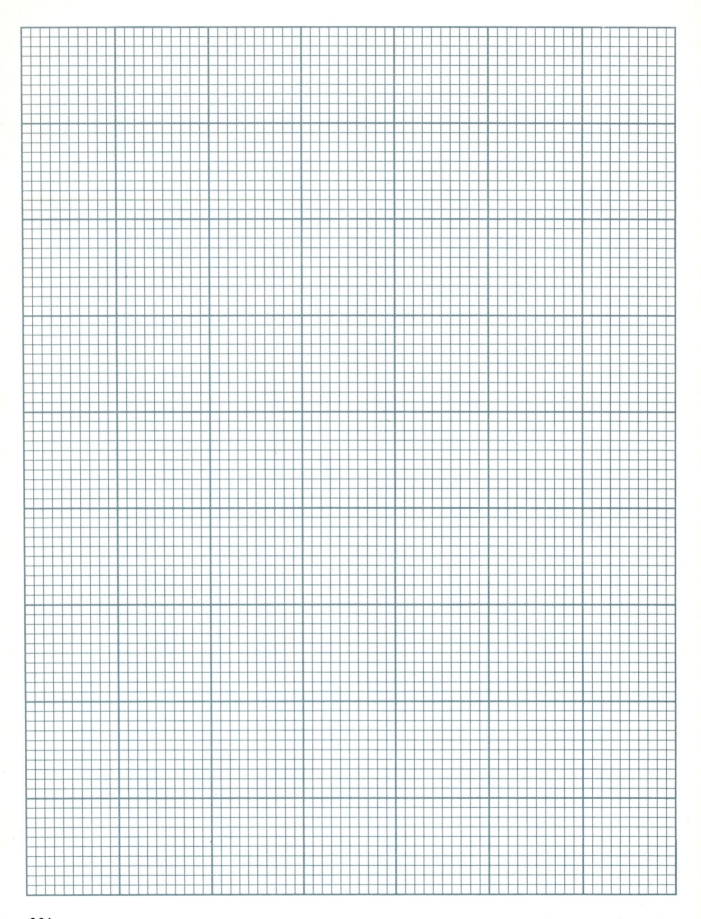

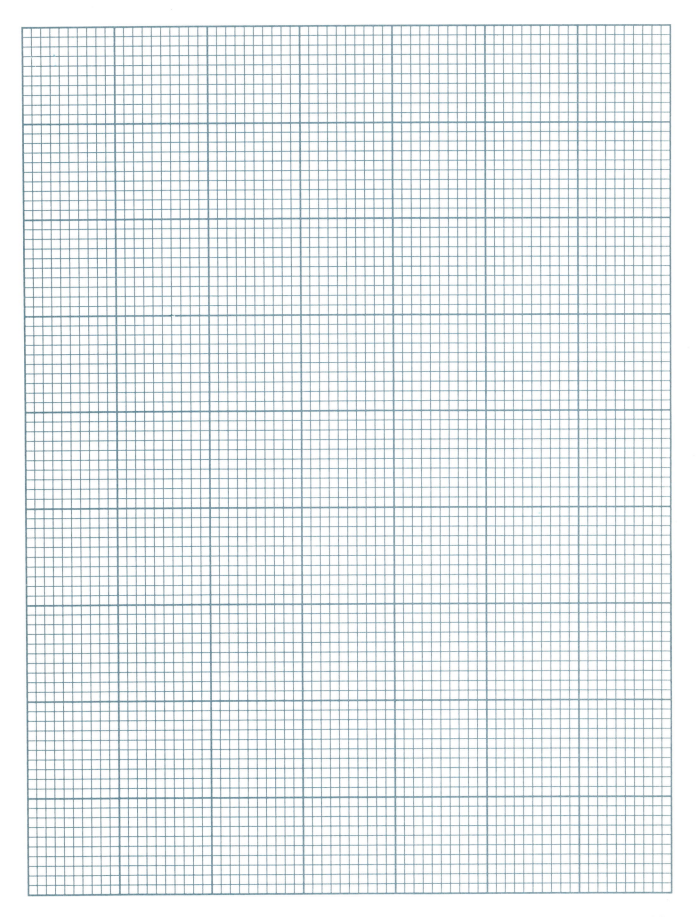

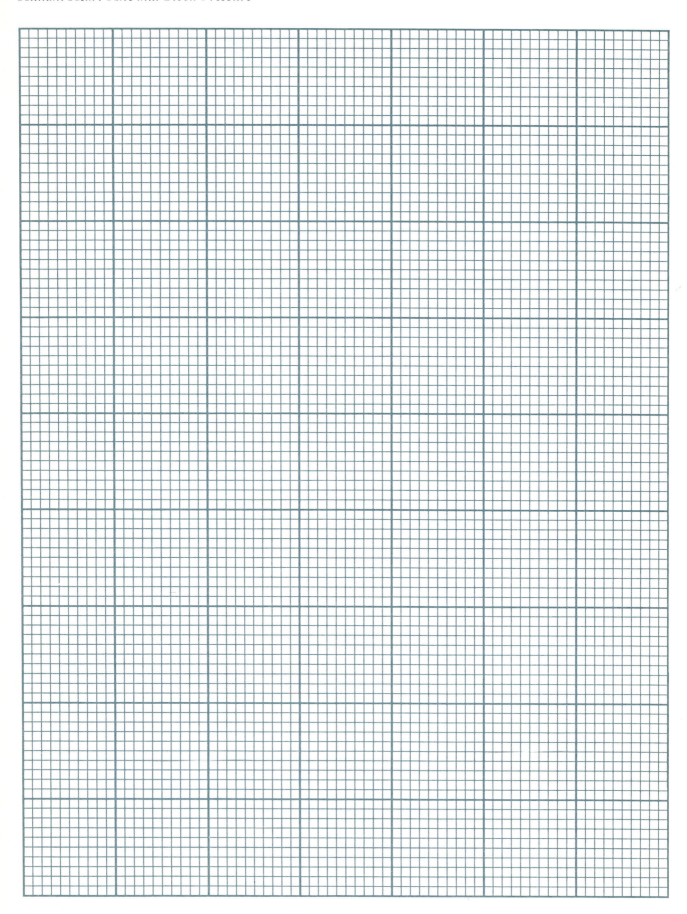

# Homeostatic Role of the Kidney

*Reading assignment: text 474-502, 511-514, 519-526*

## Objectives

### Experimental

1. To determine a) the volume of urine produced, b) urine pH, c) urine specific gravity, and d) urine $Na^+$ concentration under the following conditions:

> 1) Control,
> 2) Changes in blood volume,
> 3) Changes in salt intake,
> 4) Consumption of the base, sodium bicarbonate,
> 5) Consumption of the diuretic drug, caffeine.

### Conceptual

After completing this exercise and the reading assignment, you should be able to:

1. Describe or sketch the location of each of the following structures: **glomerulus, Bowman's capsule, proximal convoluted tubule, loop of Henle (descending limb and ascending limb), distal convoluted tubule, collecting duct, cortex of the kidney, medulla of the kidney, cortical nephron,** and **juxtamedullary nephron.**

2. Define the terms **filtrate** and **glomerular filtration rate**.

3. Describe the response of the kidney to changes in plasma volume, salt concentration, and pH.

4. Describe the mechanisms which bring about homeostatic responses to changes in plasma volume, salt concentration, and pH.

5. State the conditions under which each of the following hormones are produced: a) atrial natriuretic hormone (atriopeptin), b) antidiuretic hormone, and c) aldosterone and the organ that produces each hormone.

6. Describe the effect of caffeine on the kidney and the mechanism by which that effect is produced.

7. Describe the functions of a) a **urinometer** and b) a **flame photometer**.

## Background

The excretory system consists of the **kidneys**, the **ureters**, which connect the kidneys to the urinary **bladder**, and the **urethra**, which connects the bladder to the outside of the body. The excretory organs play a major role in maintaining steady-state solute and water concentrations. In this exercise you will have an opportunity to observe this homeostatic role of the kidneys (Fig. 19-1). You will investigate the response of the kidney to changes in the volume, salt concentration, and acidity of the blood and to a substance that increases urine output (a **diuretic**).

**Figure 19-1. A** *The anatomy of the excretory system.* **B** *An X-ray pyelogram made by passing contrast medium from the bladder up the ureters to the pelvis. The contrast medium absorbs X-rays better than the tissues do, so it outlines the hollow, fluid-filled parts of the excretory tract.*

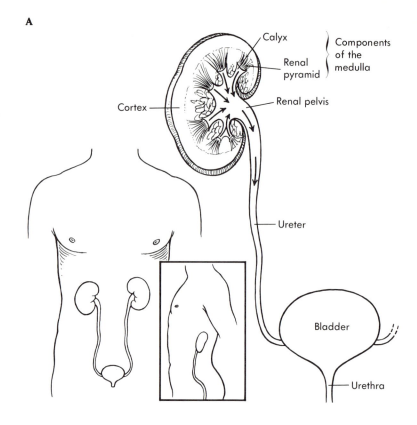

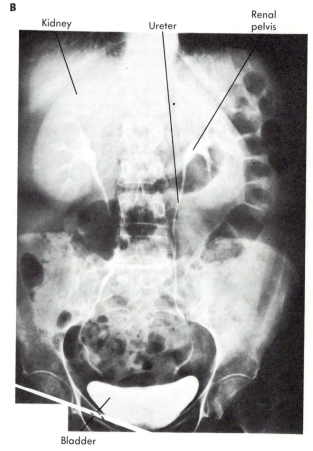

The experiments you will do involve five **treatments** in which members of the different **treatment groups** will drink different solutions (water, an isosmotic solution of NaCl, a hyperosmotic solution of NaCl, a basic solution containing sodium bicarbonate, and coffee containing the diuretic drug caffeine). Members of a **control group** will drink water only when thirsty. To determine the effects of these treatments, you will measure changes the properties of the urine produced in response to the experimental treatments during a two-hour period and compare these to the same characteristics of urine produced by the control group. You will monitor changes in four **experimental variables**:

** **volume** of urine,

** **acidity** of urine,

A

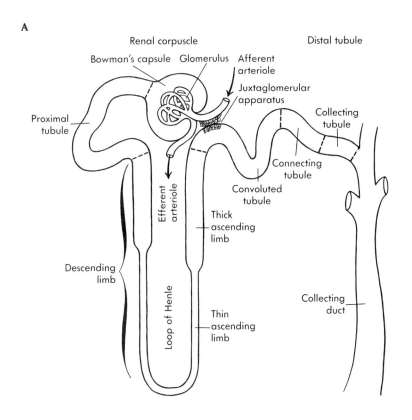

Renal corpuscle

Bowman's capsule  Glomerulus  Afferent arteriole

Distal tubule

Juxtaglomerular apparatus

Proximal tubule

Collecting tubule

Connecting tubule

Convoluted tubule

Efferent arteriole

Thick ascending limb

Descending limb

Loop of Henle

Thin ascending limb

Collecting duct

*Figure 19-2.* **A** *The functional anatomy of a nephron.* **B** *A scanning electron micrograph of a single nephron.*

B

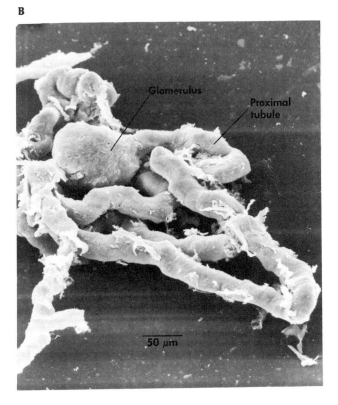

Glomerulus

Proximal tubule

50 μm

** **specific gravity** of urine, and

** **Na$^+$ concentration** of urine.

The functional unit of the kidney is the **nephron** (Fig. 19-2). At the head of each nephron is a tuft of capillaries known as the **glomerulus**, the site of plasma filtration. This capillary bed is enclosed by the funnel-shaped **Bowman's capsule**. After leaving Bowman's capsule, the tubular fluid (**filtrate**) passes through the **proximal convoluted tubule, loop of Henle (descending limb** and **ascending limb), distal convoluted tubule,** and **collecting duct**. The rate at which plasma is filtered into the glomerulus is the **glomerular filtration rate**, usually expressed in ml/min. **Cortical nephrons** are located almost entirely within the **cortex** of the kidney, while **juxtamedullary nephrons** have long loops of Henle that descend into the renal **medulla** (Fig. 19-3).

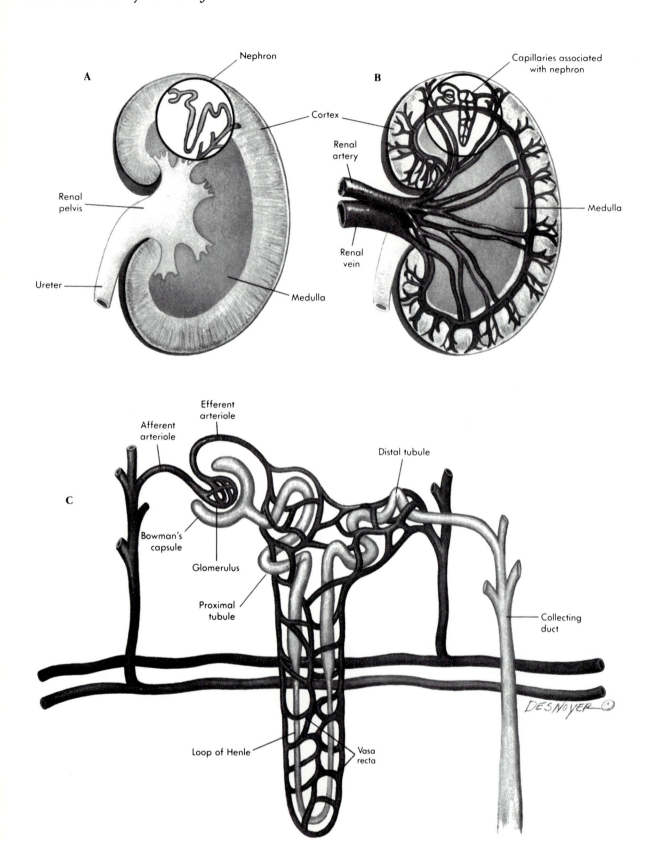

**Figure 19-3.** *A Location of a juxtamedullary nephron in the kidney. B Relationship of its glomerular and peritubular capillaries to the renal blood vessels. C An enlarged view of the nephron showing afferent and efferent arterioles, capillary beds, and venules.*

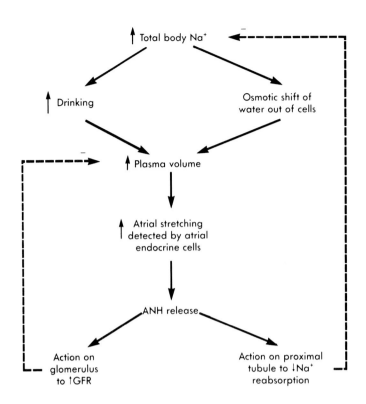

*Figure 19-4. The sequence of events in the response of the atrial natriuretic hormone (ANH) system to an increase in total body Na⁺. The feedback loop is completed when the decreased reabsorption of Na⁺ and volume changes triggered by the hormone restore extracellular fluid volume to normal. <u>GFR</u> = glomerular filtration rate.*

The regulation of salt and water balance by the kidney involves several negative feedback mechanisms, which are summarized below.

**\*\* Atrial natriuretic hormone (ANH)**, also referred to as **atriopeptin (AP)**, secreted by the atria in response to increased atrial stretch, causes increased glomerular filtration rate and Na⁺ and water loss (Fig. 19-4).

**\*\* Antidiuretic hormone (ADH)**, secreted by the posterior pituitary in response to a decrease in plasma volume or an increase in plasma osmolarity, causes increased water reabsorption across the walls of the collecting duct of the kidney, thereby decreasing urinary water loss and resulting in the production of hyperosmotic urine (Fig. 19-5). Caffeine acts by inhibiting the secretion of antidiuretic hormone (Fig. 19-6).

**\*\* Aldosterone**, secreted by the adrenal cortex in response to a regulatory cascade triggered by decreases in plasma Na⁺ concentration or plasma volume, causes increased recovery of Na⁺ in the distal tubule (Fig. 19-7).

The kidneys, along with the respiratory system, play a major role in regulating the body's acid-base balance. Under normal conditions, almost all of the bicarbonate ion (HCO₃⁻) in the filtrate is reabsorbed. However, when a large amount of base is ingested, plasma pH increases, triggering coordinated responses in the respiratory and

*Figure 19-5. The response of the antidiuretic hormone (ADH) system to a decrease in plasma volume or to an increase in plasma osmolarity is an increase in ADH secretion. The negative feedback pathways show the corrective effect of increased renal water reabsorption and increased water consumption on plasma volume and osmolarity.*

*Figure 19-6. Sites of action of diuretics.*

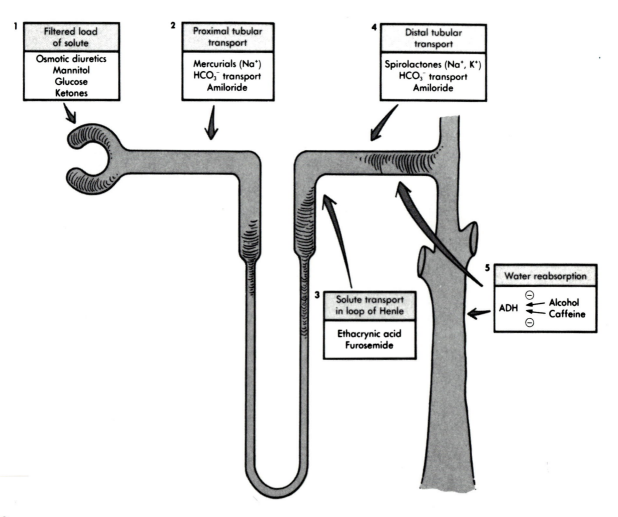

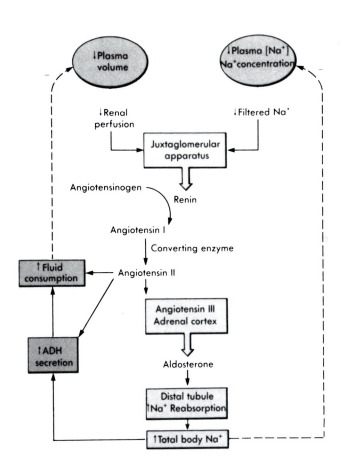

**Figure 19-7.** *The response of the aldosterone system to a decrease in plasma volume or in plasma Na⁺ concentration is an increase in renin secretion, which leads ultimately to an increase in aldosterone secretion. The increased recovery of filtered Na⁺, along with continued Na⁺ intake from diet, increases total body Na⁺. The dashed lines show the corrective effect of aldosterone secretion. If the initial problem were a volume decrease, the osmoreceptors of the ADH system would ensure that increases in total body Na⁺ resulted in restoration of volume. <u>ADH</u> = antidiuretic hormone.*

excretory systems. The immediate response is a decrease in respiration, known as **respiratory compensation**, which leads to an increase in plasma $P_{CO_2}$. **Renal compensation**, in which part of the filtered load of $HCO_3^-$ in the filtrate is not reabsorbed and some $HCO_3^-$ spills into the urine, occurs more slowly.

To be sure you understand the mechanisms you will be studying in this exercise, complete the predictions below:

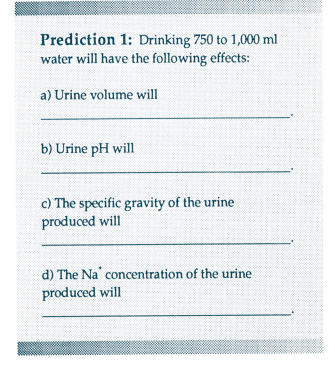

**Prediction 1:** Drinking 750 to 1,000 ml water will have the following effects:

a) Urine volume will

_____.

b) Urine pH will

_____.

c) The specific gravity of the urine produced will

_____.

d) The Na⁺ concentration of the urine produced will

_____.

**Prediction 2:** Drinking 750 to 1,000 ml 0.9% NaCl will have the following effects:

a) Urine volume will
_____ .

b) Urine pH will
_____ .

c) The specific gravity of the urine produced will
_____ .

d) The $Na^+$ concentration of the urine produced will
_____ .

**Prediction 4:** Drinking 500 ml sodium bicarbonate will have the following effects:

a) Urine volume will
_____ .

b) Urine pH will
_____ .

c) The specific gravity of the urine produced will
_____ .

d) The $Na^+$ concentration of the urine produced will
_____ .

**Prediction 3:** Drinking 75 to 100 ml 5% NaCl solution will have the following effects:

a) Urine volume will
_____ .

b) Urine pH will
_____ .

c) The specific gravity of the urine produced will
_____ .

d) The $Na^+$ concentration of the urine produced will
_____ .

**Prediction 5:** Drinking 250 to 500 ml coffee will have the following effects:

a) Urine volume will
_____ .

b) Urine pH will
_____ .

c) The specific gravity of the urine produced will
_____ .

d) The $Na^+$ concentration of the urine produced will
_____ .

## Procedure

Record your results in the laboratory report section at the end of this exercise.

Caution: some individuals should not take caffeine and others should avoid salt. If you are in one of these categories, inform your instructor, so that you can be assigned to the control group.

1. The class will be divided into five approximately equal treatment groups. Assemble with your group and begin immediately. Each group will measure the effects of a different treatment on the volume, acidity, total solute concentration, and $Na^+$ concentration of urine. One group will drink only when thirsty; the other five groups will drink various test solutions. To save time and reduce the number of determinations to be made, all tests will be made on urine pooled from all members of a treatment group. This is like treating each group as a single person.

2. Before you begin this exercise, be sure you are familiar with the test procedure.

3. Collect the contents of your bladder. This is your control sample. Each member of the group should return to the laboratory from the restroom with a sample.

4. Note the time in Table 19-1. If you are in one of the treatment groups, drink the test solution for your group (see list below), as rapidly as is comfortable:

Control group:

> Group A. Do not drink a test solution. Drink water only when thirsty, just as you normally would.

Treatment groups:

> Group B. Each person drinks 750 to 1,000 ml distilled water.

> Group C. Each person drinks 750 to 1,000 ml 0.9% NaCl solution.

> Group D. Each person drinks 75 to 100 ml 5% NaCl solution. This will be a difficult solution to get down, so don't gulp it. Eating pretzels may help. If you begin to feel nauseated, stop drinking at that point and drink a little water.

> Group E. Each person drinks 500 ml 3% sodium bicarbonate solution.

> Group F. Each person drinks 250 to 500 ml coffee. Members of this group should not have had any coffee during the day.

5. Thirty min after collecting your initial sample, collect another sample. This is your first experimental sample. Continue to collect and test samples at 30 min intervals until the end of the laboratory period. You should be able to collect and analyze a total of five samples during the class period.

6. In the intervals between collecting samples, determine volume, pH, specific gravity, and $Na^+$ concentration, using the procedures described below. Enter all results in Table 19-1.

a) Volume

The volume of your sample can be read using the graduations on the beaker it is collected in. Add this figure to that of the other group members and divide by the number of members to determine average volume.

Use the 25 ml graduated cylinder to measure out 25 ml of your urine. Add this to the beaker containing the pooled sample. When all members have added their 25 ml, swirl the solution to mix. The pH, specific gravity, and $Na^+$ concentrations of this pooled sample will be determined. Discard the rest.

## b) pH

To measure the pH of your pooled sample, dip the end of a strip of pH paper into it. Only about 0.5 cm of paper should be dipped for each determination. Compare the color of the freshly dipped strip with the color guide included with the paper. The pH paper can be used only once, so tear off and discard dipped portions of the paper when you are finished with them.

## c) Specific gravity

The specific gravity of a solution is its density compared to that of pure water. This is a measure of the total amount of solute dissolved in the solution. Because specific gravity is a ratio of two densities, the units in the numerator and the denominator cancel out. Hence, specific gravity is a dimensionless number. The specific gravity of water is approximately 1.0.

To measure the specific gravity of a urine sample, fill the underlined urinometer tube with enough urine for the float to be able to bob freely without touching the bottom. To be sure the float is not clinging to the side of the tube, spin it gently. When it stops spinning, note the line on the float calibration that is exactly at the surface of the fluid. The float floats higher in solutions which contain higher concentrations of solute.

## d) $Na^+$ concentration

Measure 1 ml of pooled urine into the 200 ml volumetric flask. Your instructor will demonstrate the use of the pipette bulb. To use the 1 ml pipette accurately, drain its contents into the volumetric flask. Some urine will remain in the tip. Do not blow or shake this into the volumetric flask. Do not attempt to rinse the pipette, since the small amount of urine that clings to the sides of the pipette will introduce less error into the $Na^+$ determination than the same amount of distilled water. After you drain the 1 ml pipette into the volumetric flask used in preparing your diluted urine, dispose of the tip.

Fill the flask to the mark with distilled water. Mix the solution by shaking the flask. When it is mixed, fill a clean test tube with this diluted urine. Label it with your group number and sample number. Your instructor will determine its $Na^+$ concentration using a flame photometer. The units for this measurement are milliequivalents per liter (mEq/l).

7. Clean your glassware. After you have determined volume, pH, and specific gravity and prepared your diluted urine sample, you need to ready your glassware for the next sample, as follows:

> The collection beaker should be rinsed with distilled or deionized water and wiped dry.
>
> The graduated cylinder should be drained thoroughly but need not be rinsed.
>
> The urinometer float should be rinsed with distilled or deionized water and wiped off. Be careful, since it is fragile.
>
> Drain the urinometer cylinder thoroughly.
>
> Rinse the volumetric flask with distilled or deionized water. Do not worry if some drops of rinse water remain on the walls of the flask.

# Laboratory Report

*Exercise 19: Homeostatic Role of the Kidney*

Name: _____

Date: _____

Lab Section: _____

## Analyzing Your Data

Table 19-1 summarizes your data on changes in urine volume, pH, specific gravity, and $Na^+$ concentration. Data from all groups will be pooled in tables 19-2 to 19-5. Enter your group's results on the board and copy down the values for each of the other groups, so that you will have a complete data set for all variables and groups.

Make four graphs showing changes in the four experimental variables over time. (For example,

*Table 19-1. Characteristics of urine produced during 2-hr experiment.*

| Treatment | | | Starting time: | | |
|---|---|---|---|---|---|
| | | | | | |
| | | Time (min) from start of experiment | | | |
| Urine characteristics | 0[1] | 30 | 60 | 90 | 120 |
| Average volume (ml) | | | | | |
| pH | | | | | |
| Specific gravity | | | | | |
| $Na^+$ concentration (mEq/l) | | | | | |

[1] Control sample.

297

your first graph should show urine volume as a function of time, and your second graph will show changes in urine pH as a function of time.) Each graph will need six separate lines to show the trends for the five treatment groups plus the control group. Be sure these are clearly labeled.

Did your data conform to your expectations? If not, was this because of something you didn't take into consideration when you formulated your predictions? Or was it because of something unusual about your data?

*Table 19-2. Changes in urine volume.*

| | Volume of urine produced (ml) | | | | |
|---|---|---|---|---|---|
| | Time (min) from start of experiment | | | | |
| Treatment | $0^1$ | 30 | 60 | 90 | 120 |
| Water as needed | | | | | |
| 750-1,000 ml water | | | | | |
| 750-1,000 ml 0.9% NaCl | | | | | |
| 75-100 ml 5% NaCl | | | | | |
| 500 ml 0.3% $NaHCO_3$ | | | | | |
| 250-500 ml coffee | | | | | |

[1] Control sample.

*Table 19-3. Changes in urine pH.*

| | pH of urine produced | | | | |
|---|---|---|---|---|---|
| | Time (min) from start of experiment | | | | |
| Treatment | 0[1] | 30 | 60 | 90 | 120 |
| Water as needed | | | | | |
| 750–1,000 ml water | | | | | |
| 750–1,000 ml 0.9% NaCl | | | | | |
| 75–100 ml 5% NaCl | | | | | |
| 500 ml 0.3% $NaHCO_3$ | | | | | |
| 250–500 ml coffee | | | | | |

[1] Control sample.

*Table 19-4. Changes in specific gravity of urine.*

| | Specific gravity of urine produced | | | | |
|---|---|---|---|---|---|
| | Time (min) from start of experiment | | | | |
| **Treatment** | **0[1]** | **30** | **60** | **90** | **120** |
| **Water as needed** | | | | | |
| 750-1,000 ml water | | | | | |
| 750-1,000 ml 0.9% NaCl | | | | | |
| 75-100 ml 5% NaCl | | | | | |
| 500 ml 0.3% NaHCO$_3$ | | | | | |
| 250-500 ml coffee | | | | | |

[1] Control sample.

*Table 19-5. Changes in Na$^+$ concentration of urine.*

| Treatment | Na$^+$ concentration of urine produced (mEq/l) | | | | |
|---|---|---|---|---|---|
| | Time (min) from start of experiment | | | | |
| | 0[1] | 30 | 60 | 90 | 120 |
| Water as needed | | | | | |
| 750-1,000 ml water | | | | | |
| 750-1,000 ml 0.9% NaCl | | | | | |
| 75-100 ml 5% NaCl | | | | | |
| 500 ml 0.3% NaHCO$_3$ | | | | | |
| 250-500 ml coffee | | | | | |

[1] Control sample.

## Questions

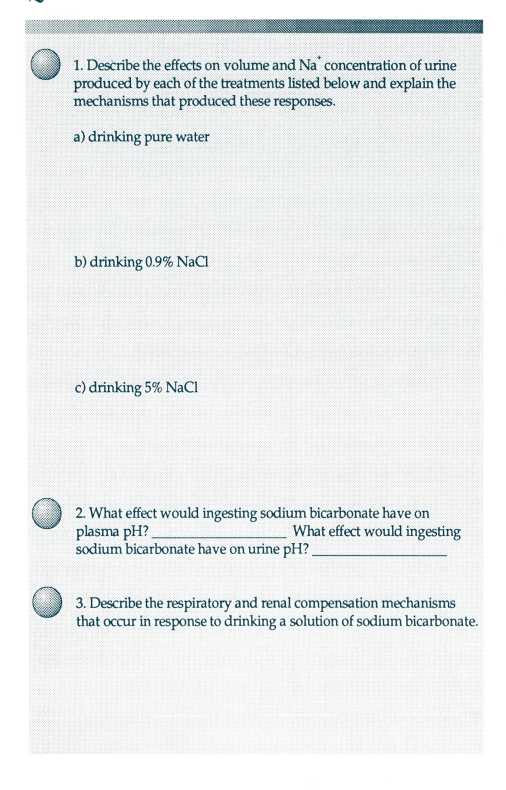

1. Describe the effects on volume and $Na^+$ concentration of urine produced by each of the treatments listed below and explain the mechanisms that produced these responses.

a) drinking pure water

b) drinking 0.9% NaCl

c) drinking 5% NaCl

2. What effect would ingesting sodium bicarbonate have on plasma pH? _____ What effect would ingesting sodium bicarbonate have on urine pH? _____

3. Describe the respiratory and renal compensation mechanisms that occur in response to drinking a solution of sodium bicarbonate.

4. In the group that drank coffee, did urine specific gravity decrease or increase? _____ In the group that drank coffee, did the Na$^+$ concentration of urine decrease or increase? _____

5. Describe the effects of caffeine on the kidney that produced the changes you observed.

6. Describe the effects of each of the hormones listed below:

a) atrial natriuretic hormone (atriopeptin)

b) antidiuretic hormone

c) aldosterone

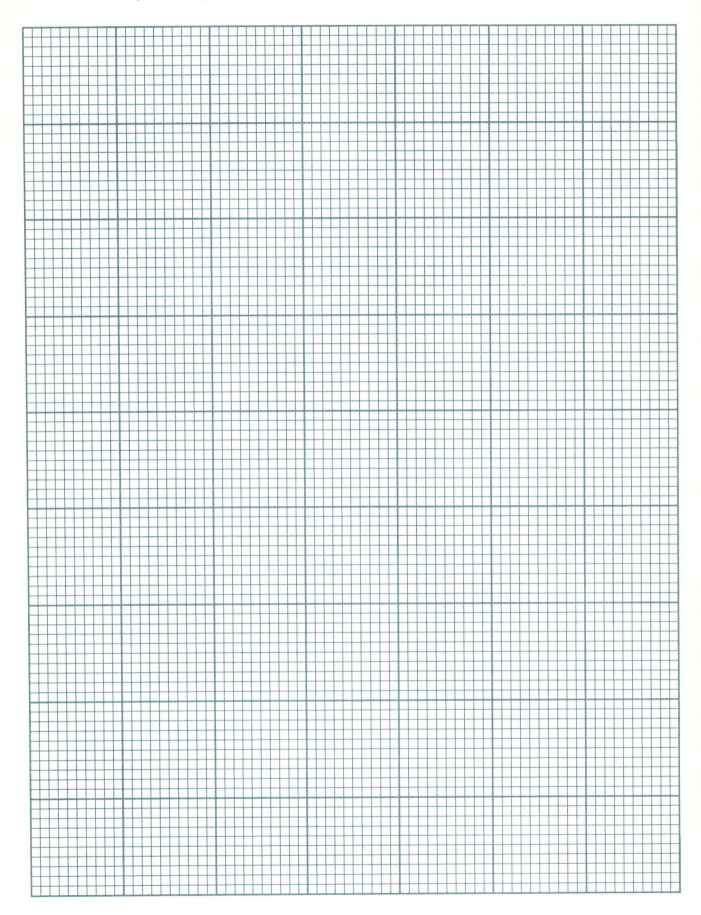

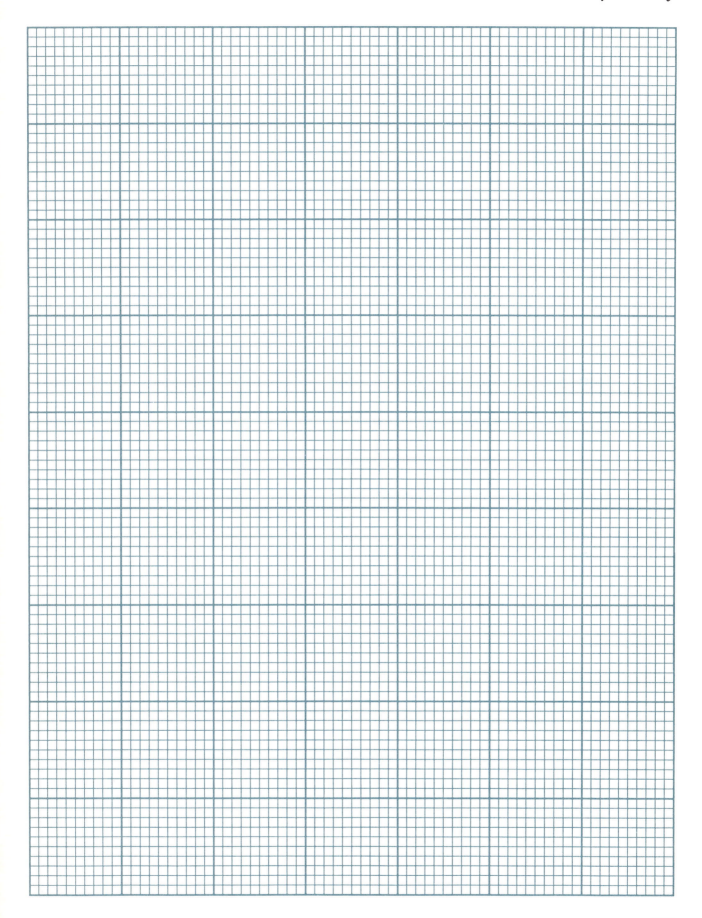

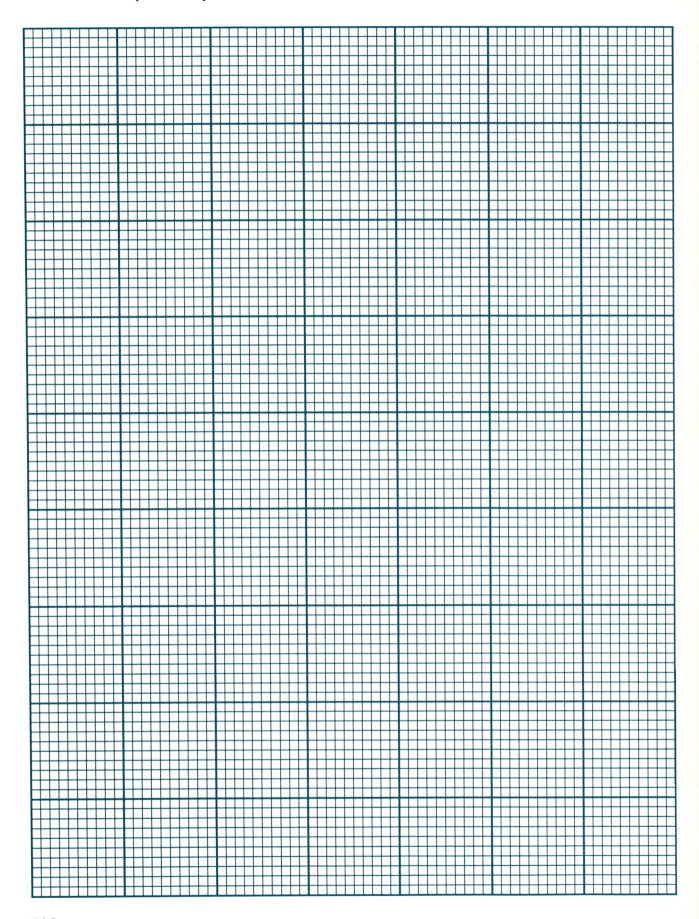

# 20 Effects of Digestive Secretions

*Reading assignment: text 40-50, 567-573*

## Objectives

### Experimental

1. To determine the effects of pH on the hydrolysis of starch by salivary amylase.

2. To determine the effects of rennin on milk protein.

3. To determine the effects of temperature and pH on the hydrolysis of albumen by pepsin.

4. To determine the effects of pancreatic enzymes on proteins, carbohydrates, and fats.

### Conceptual

After completing this exercise and the reading assignment, you should be able to:

1. Describe the structure of carbohydrates, fats, and proteins.

2. Describe the locations where enzymatic digestion of carbohydrates, fats, and proteins takes place.

3. List enzymes that catalyze the hydrolysis of carbohydrates, fats, and proteins.

4. Name the products of enzymatic digestion of carbohydrates, fats, and proteins.

5. Define **emulsification** and name the substance that is responsible for emulsification of fats in the body.

6. Explain why fats, but not proteins or carbohydrates, must be emulsified before substantial digestion can take place.

7. Describe the effects of a) pH on salivary amylase, b) rennin on casein, c) temperature and pH on pepsin, and d) pancreatic enzymes on carbohydrates, fats, and proteins.

8. Describe one method for detecting each of the following: a) hydrolysis of starch, b) hydrolysis of protein, and c) hydrolysis of fats.

## Background

Most of our food consists of three classes of **macromolecules**: carbohydrates, proteins, and fats. Sugars, the smallest of these compounds, have molecular weights in the hundreds, while molecular weights of some proteins are in the millions. If foodstuffs are to be used, either for fuel or for synthesis, the large molecules we ingest must be broken down into smaller molecules before they can be incorporated into the body or metabolized further. **Digestion**, the process of breaking down large and complex food molecules into smaller units that can be absorbed, occurs in the digestive tract, with the help of enzymes.

**Carbohydrates** are organic molecules composed of carbon, hydrogen, and oxygen in the ratio of 1:2:1. Sugars and starches are included in this class of organic compounds. Starches (Fig. 20-1)

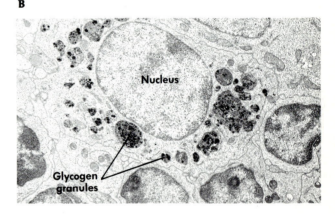

Branch

Main chain

*Figure 20-1.* **A** *Structure of glycogen (animal starch) showing one of the characteristic branches from the main chain.* **B** *An electron micrograph of glycogen granules in a liver cell.*

are polymers of glucose. When they are broken down, starches first yield smaller **polysaccharides**, then **disaccharide** subunits (Fig. 20-2), and, ultimately, **monosaccharides**.

Starch digestion begins in the mouth when starch is mixed with saliva containing the enzyme **salivary amylase**. Salivary amylase digests about half the starch you eat; the rest is digested in the **duodenum** by pancreatic amylase. It is no accident that many human cultures prepare starchy foods, such as potatoes and grains, by cooking them, for amylase breaks starch down more readily if the starch has been heated first. Animals that digest uncooked starch may be aided by bacteria in the initial stages of starch digestion.

**Proteins** are composed of one or more **polypeptide** chains, which are assembled by the formation of **peptide bonds** between the **carboxylic acid** group of one amino acid and the **amino** group of a second amino acid (Fig. 20-3). Digestion of proteins begins in the stomach with the hydrolysis of large polypeptides into smaller polypeptides by a class of enzymes termed **pepsin**. Further digestion of proteins takes place in the small intestine and involves the enzymes **trypsin** and **chymotrypsin**, secreted by the pancreas.

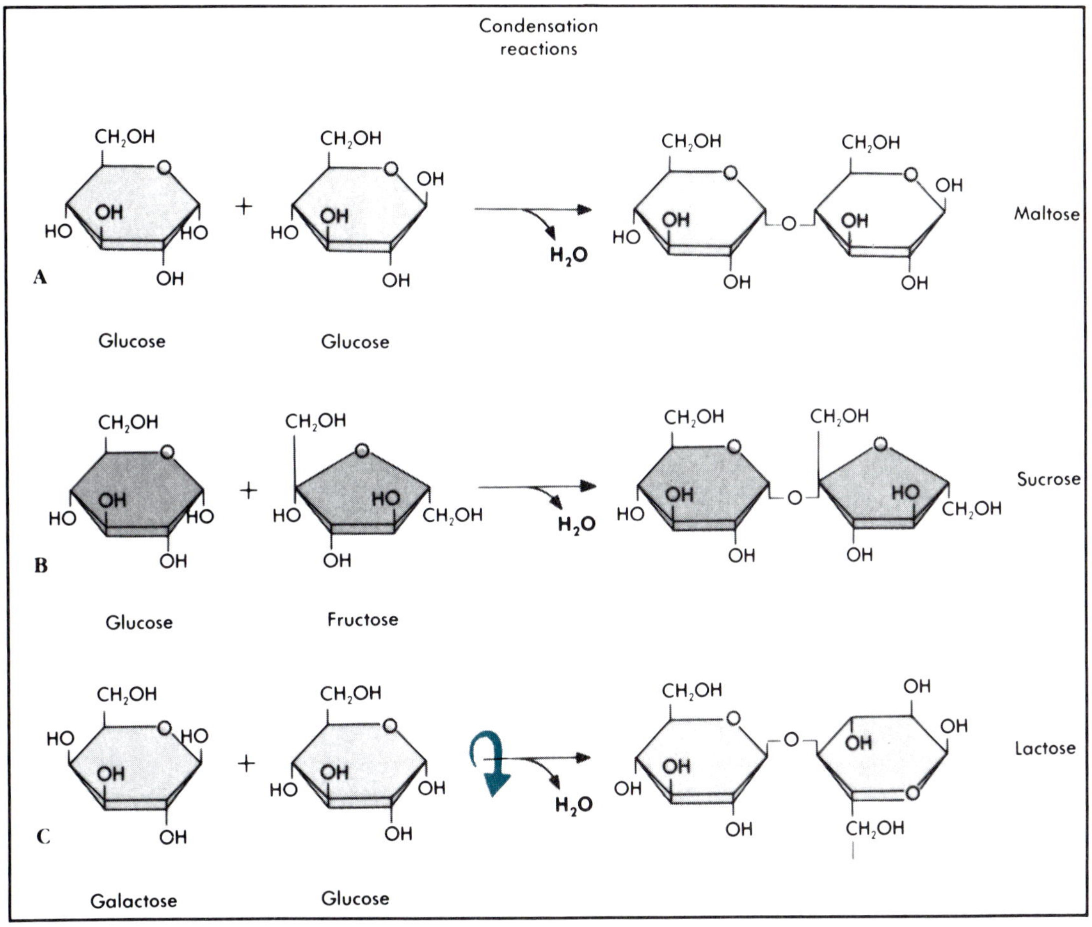

*Figure 20-2. Some examples of disaccharides. A Apples are rich in maltose, composed of two glucose molecules. B In table sugar, glucose and fructose are joined to form sucrose. C Milk contains a disaccharide of glucose and galactose termed lactose or milk sugar. Blue-green arrow indicates that one of the subunit molecules of lactose is inverted relative to the other, when galactose and glucose are joined. As a a result of this inversion, lactose contains different bonds than maltose and sucrose, a property that makes milk difficult to digest for individuals who lack the enzyme, lactase.*

Compounds composed largely of hydrocarbons that usually are not water soluble are termed **lipids (fats)**. Most dietary fat consists of **triglycerides** (neutral fats), **phospholipids, fatty acids**, and **cholesterol** or **cholesterol esters** (Fig. 20-4). Hydrolysis of fats produces **glycerol** and fatty acids. Because fats are not soluble in water, the digestion of fats is much more efficient if they first undergo **emulsification** (Fig. 20-5). In this process, bile salts produced in the liver act like a detergent, lowering the surface tension of fat droplets, and allowing them to break up into

smaller droplets. This increases the surface area of lipids available to digestive enzymes. The digestion of fats occurs primarily in the small intestine, in the presence of **pancreatic lipase**.

In this exercise, you will observe enzymatic digestion of carbohydrates, fats, and proteins. To do this you will expose sample foods to various digestive conditions in the sequence in which the body digests food: mouth, stomach, and small intestine. The foods you will use are:

*Figure 20-3. Amino acids combine by way of peptide bonds to form polypeptide chains. When a peptide bond is formed, the carboxylic acid (COOH) group of one amino acid combines with the amino (NH₂) group of another amino acid, with the removal of a molecule of water.*

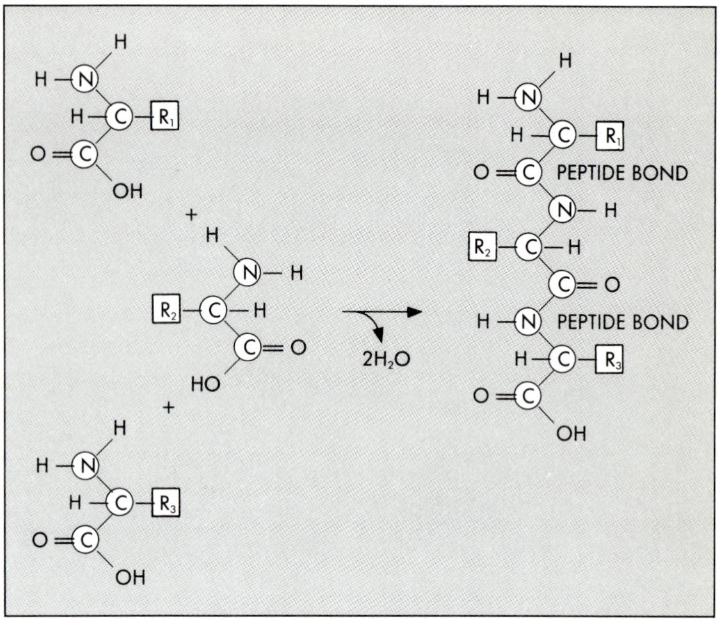

1. A solution of cooked **starch**,

2. **Skimmed milk** (containing a mixture of compounds, including the milk protein, **casein**),

3. Boiled **egg white** (containing the protein **albumen**), and

4. **Cream** (which is high in fat).

These foods will be exposed to the following enzymes:

1. **Salivary amylase**: the enzyme in saliva that breaks carbohydrates down into sugars. (This enzyme will be obtained by collecting and pooling saliva from all members of the class, because diet and time of day influence the production of salivary amylase, and some people do not produce any salivary amylase.)

2. **Rennin: a coagulating enzyme found in the stomachs of young mammals (and to a much lesser extent in the stomach of some adults) that coagulates casein and causes it to remain in the stomach long enough to be acted upon by pepsin.**

3. **Pepsin**: a class of protein-digesting enzymes secreted by the stomach.

4. **Pancreatin**: a commercial preparation of pancreas extract containing many enzymes that are secreted by the pancreas and act on carbohydrates, proteins, and fats in the small intestine.

In Part 1 you will observe hydrolysis of starch by salivary amylase. You will use **I₂-KI** to test for the presence of substrate (starch). Starch turns dark blue in the presence of iodine. Therefore, when starch is mixed with I₂-KI, the development of a blue-black color indicates that starch has **not** been hydrolyzed. If a solution of partially hydrolyzed starch is tested with I₂-KI, a red color will develop. When the starch has been completely digested, it will be colorless in the presence of I₂-KI, and the test solution will appear straw-colored (the color of the I₂-KI).

In parts 2 and 3 you will observe protein digestion. You will determine the effects of rennin on milk protein (Part 2) and the effects of pH and temperature on hydrolysis of the protein albumen by pepsin (Part 3). **Biuret reagent** combines with peptide bonds to produce a violet color. Therefore, you should expect to observe a bright, violet color

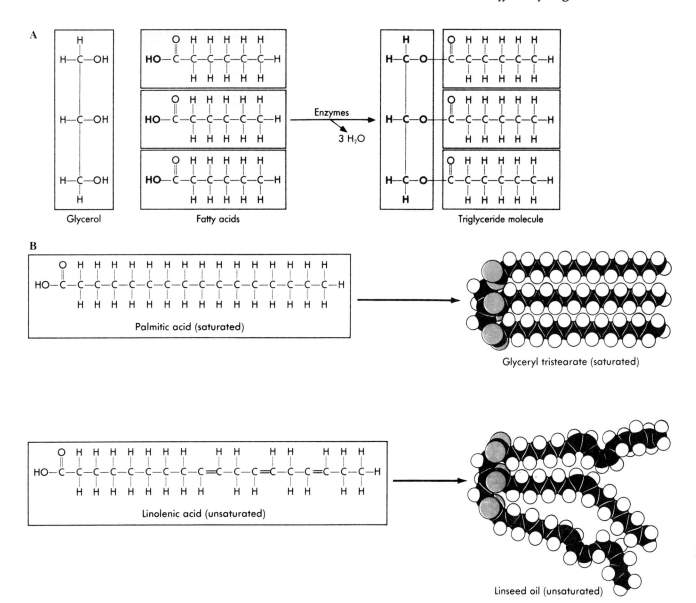

*Figure 20-4.* **A** *Triglycerides are molecules composed of three fatty acids coupled to a glycerol backbone.* **B** *Fatty acids are saturated, if they lack double bonds in the hydrocarbon tail (palmitic acid), or unsaturated. if they contain one or more double bonds (linolenic acid) Many triglycerides from animals are saturated. Because their fatty acid chains fit closely together, these triglycerides form immobile arrays called hard fat. However, many vegetable oils, such as linseed oil, are unsaturated, and their double bonds prevent close association of the triglycerides.*

at the outset of the experiment, and the color should diminish as protein digestion proceeds and peptide bonds are broken.

In Part 4 you will investigate the effect of pancreatic enzymes on starch, protein, and fat. Because hydrolysis of fats yields fatty acids and glycerol, the digestion of fat is accompanied by an increase in free fatty acids and a decrease in pH. Therefore, you will follow the progress of fat digestion by monitoring pH. **Litmus cream** contains an indicator that is blue in alkaline conditions and red in acidic solutions.

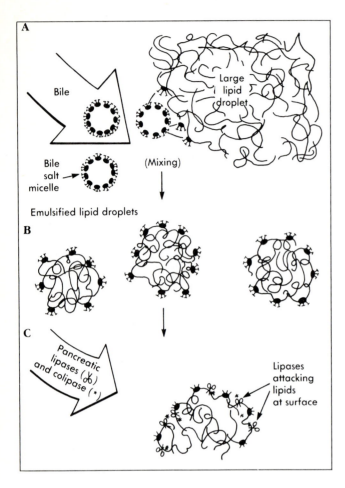

*Figure 20-5. The emulsification and digestion of lipids.*
*A Bile salts enter the lumen of the gastrointestinal tract*
*in the form of spherical micelles, with polar groups*
*extending toward the surrounding water and hydrophobic*
*regions oriented toward the center of the micelle. Mixing*
*of the intestinal contents brings the bile salt micelles into*
*contact with large aggregates of lipids. B Some of the*
*lipids are separated from the large masses when their*
*surfaces become covered with bile salts. C Pancreatic*
*lipases, with the help of a pancreatic polypeptide known*
*as colipase, attack the bonds in lipids exposed on the*
*surface of the emulsified lipid droplet.*

## Procedure

Record your results in the laboratory report section at the end of this exercise.

 **Part 1. Starch digestion in the mouth**

1.  After rinsing your mouth with water, collect about 2 cm of saliva in a test tube. (If desired, you can chew a piece of paraffin or a rubber band to aid salivation.)

2. Dilute the saliva in the test tube with an equal volume of water. Add your diluted saliva to the beaker containing the pooled class sample. Your instructor will strain this dilute enzyme mixture through cheesecloth.

3. Obtain about 10 ml of the diluted, pooled enzyme mixture. Using a glass rod, remove a drop of enzyme solution and place it on a piece of pH paper. Record the pH of this solution in Table 20-1.

4. Label 6 clean test tubes, 1 to 6. Add 5 ml of boiled starch solution to each test tube (Fig. 20-6).

5. To prepare starch-buffer mixtures, mark the tubes as follows:

> Tube 1: pH 5,
> Tube 2: pH 6,
> Tube 3: pH 7,
> Tube 4: pH 7,
> Tube 5: pH 8,
> Tube 6: pH 9,

and add 2 ml of the appropriate buffer solution to each tube. (Note that there are 2 tubes that receive the pH 7 buffer.)

6. Prepare a spot plate for the starch test. Place 3 drops of $I_2$-KI solution in one or two of the depressions of your spot plate. (Do not put $I_2$-KI into more than two depressions at any one time, because prolonged exposure to air inactivates it.)

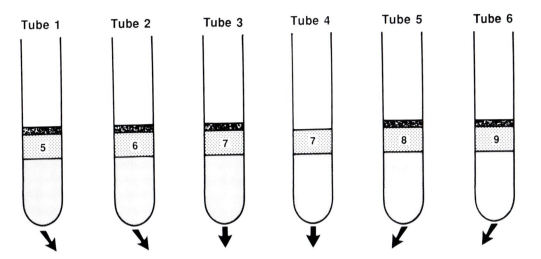

*Figure 20-6. Experimental set-up for determining effect of salivary amylase on starch.*

I₂-KI Test for Presence of Starch

- 10 drops enzyme
- 2 ml buffer (number=pH)
- 5 ml boiled starch

7. At Time 0 add 10 drops of enzyme solution to Tube 1 (pH 5). Seal the tube with parafilm and gently invert the tube to mix the contents. (<u>Do not shake the mixture vigorously</u>. Froth indicates that the enzyme has been denatured.) Record the time in Table 20-1.

8. Repeat Step 7 using Tube 6 (pH 9).

9. Test for hydrolysis of starch at pH 5, 1 min after adding enzyme to Tube 1. Remove a drop of the solution from Tube 1 and place it in the I₂-KI in one of the depressions in the spot plate. Note the color. If a dark blue color does not develop, check your procedure with your instructor before you continue. This is your starch test for Time 1. Record your results in Table 20-1.

10. Repeat Step 9 for Tube 6.

11. If a blue color develops in either tube, continue testing for starch at 1 min intervals until there is a definite change in color, indicating that starch is being hydrolyzed. Select an endpoint

color and use it for all subsequent tests. That is, if you run your first test only until a red color develops, then terminate all your subsequent tests when the same red color is observed. Record all results in Table 20-1.

Time can be saved if you test for the presence of starch only as often as necessary. If three consecutive tests show no change in color, triple the time between the tests. If color changes are evident, reduce the test interval accordingly. Once you detect that you have a slowly-reacting system, start another tube or two (see below), so that you can monitor several at the same time. If you are in doubt about the time for the reaction end point with a relatively fast reaction, repeat it to confirm your results.

12. When you have completed your test for starch hydrolysis at pH 5 and pH 9, clean your spot plate. Repeat steps 6 to 11 with tubes 2 and 5, containing buffers at pH 6 and 8. Record your results in Table 20-1. Then repeat steps 6 to 11 with tubes 3 and 4 (pH 7), but <u>do not add enzyme</u>

to <u>Tube 4</u>. In all other respects, your handling of Tube 4 should be identical to your treatment of the other tubes.

 *Part 2. Effect of rennin on milk protein*

1. Number 2 clean test tubes. Place 2 ml skimmed milk in each tube. Add additional reagents as follows (Fig. 20-7):

Tube 1: 2.5 ml 5% rennin solution,
Tube 2: 2.5 ml H$_2$0.

2. Examine the tubes for the presence of curd. Record your observations in Table 20-2.

 *Part 3. Effect of pepsin on albumen*

<u>Caution: extreme care must be used to avoid contact of skin or clothes with strong acid or base.</u>

1. Label 5 clean test tubes, 1 to 5.

2. Cut thin slices of boiled egg white. Add 1 slice to each tube.

3. Add the following reagents to each tube (Fig. 20-8) and incubate at the temperature indicated for at least 1 hr (watch the temperature carefully):

Tube 1: 1 drop distilled water + 5 ml pepsin solution, incubate at 37°C;

Tube 2: 1 drop concentrated HCl + 5 ml pepsin solution, incubate at 37°C;

Tube 3: 1 drop 10 N NaOH + 5 ml pepsin solution, incubate at 37°C;

Tube 4: 1 drop concentrated HCl + 5 ml pepsin solution, incubate at 0°C;

Tube 5: 1 drop concentrated HCl + 5 ml water, incubate at 37°C.

Shake tubes periodically during incubation.

4. To test for the presence of peptide bonds, filter each solution into a clean test tube. Add 5 drops Biuret reagent to each tube. Record the color in each tube in Table 20-3.

 *Part 4. The effect of pancreatic enzymes on macromolecules*

1. Using 2 ml starch, 1.5 ml pancreatin, and 2 ml of each of the buffers used in Part 1, repeat the test for hydrolysis of starch (Fig. 20-9). This time place the tubes in a 37°C water bath for 15 min before performing the I$_2$-KI test. Record your results in Table 20-4A.

*Figure 20-7. Experimental set-up for determining effect of rennin on milk protein.*

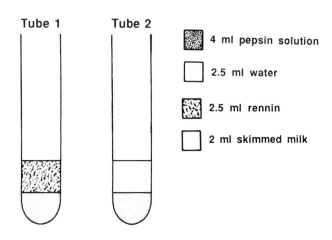

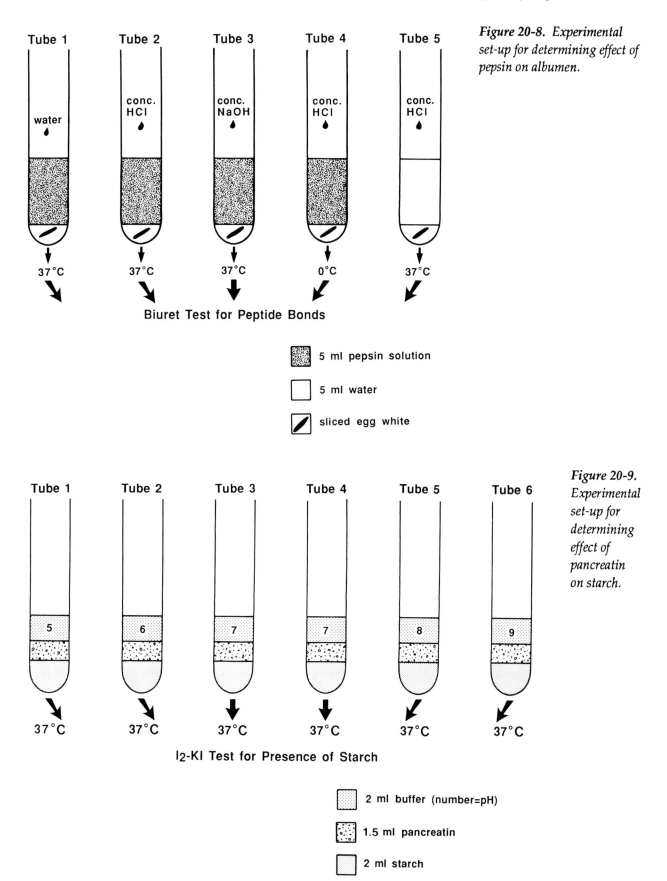

Tube 1    Tube 2    Tube 3    Tube 4    Tube 5

water    conc. HCl    conc. NaOH    conc. HCl    conc. HCl

37°C    37°C    37°C    0°C    37°C

*Figure 20-8. Experimental set-up for determining effect of pepsin on albumen.*

Biuret Test for Peptide Bonds

5 ml pepsin solution

5 ml water

sliced egg white

Tube 1    Tube 2    Tube 3    Tube 4    Tube 5    Tube 6

5    6    7    7    8    9

37°C    37°C    37°C    37°C    37°C    37°C

*Figure 20-9. Experimental set-up for determining effect of pancreatin on starch.*

$I_2$-KI Test for Presence of Starch

2 ml buffer (number=pH)

1.5 ml pancreatin

2 ml starch

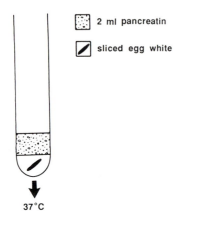

2 ml pancreatin

sliced egg white

37°C

*Figure 20-10.  Experimental set-up for determining effect of pancreatin on egg white.*

**2. Test for hydrolysis of protein (Fig. 20-10).  Place a slice of egg white and 2 ml pancreatin in a clean test tube.  Place the tube in a 37°C water bath for 1 hr.  Filter the solution into a clean test tube, add 5 drops of Biuret reagent, and record any color changes in Table 20-4B.**

**3. Test for hydrolysis of fats (Fig. 20-11).  Set up 4 test tubes as follows:**

> **Tube 1: 2 ml litmus cream + 2 ml pancreatin,**
>
> **Tube 2: 2 ml litmus cream + 2 ml distilled water,**
>
> **Tube 3: 2 ml litmus cream + 2 ml pancreatin + bile salt (just enough to cover the tip of the spatula),**
>
> **Tube 4: 2 ml litmus cream + 2 ml distilled water + bile salt (again, use only enough to cover the tip of your spatula).**

**4. Incubate tubes 1 to 4 for 1 hr.  Record any color changes you observe, in Table 20-4C.**

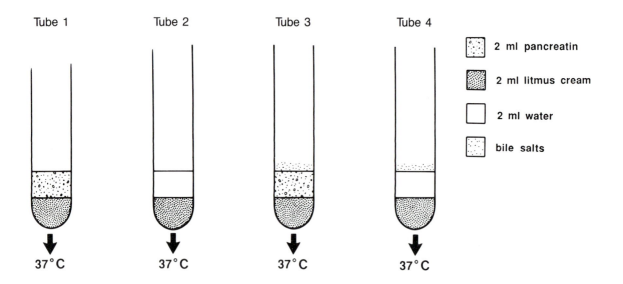

Tube 1   Tube 2   Tube 3   Tube 4

2 ml pancreatin

2 ml litmus cream

2 ml water

bile salts

37°C   37°C   37°C   37°C

*Figure 20-11. Experimental set-up for determining effect of pancreatin on fats.*

# Laboratory Report

### Exercise 20: Effects of Digestive Secretions

Name: _____

Date: _____

Lab Section: _____

## Analyzing Your Data

Plot the relationship between pH (on the $x$ axis) and the rate of starch hydrolysis ($y$ axis) for 1) salivary amylase (Table 20-1) and 2) pancreatin (Table 20-4A). Remember that rate is the reciprocal of time. Thus, a long time to hydrolysis means a slow reaction rate and vice versa.

In addition, use your observations and the data you recorded in tables 20-1 to 20-4 to answer the questions at the end of this exercise.

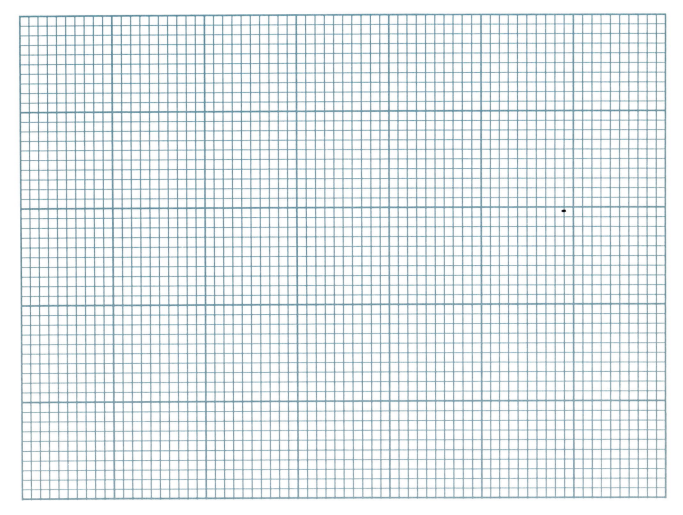

**Table 20-1.** *Effect of pH on hydrolysis of starch by salivary amylase, determined by $I_2$-KI test for starch. (+ indicates that a dark blue color was observed.)*

pH of dilute saliva: _____ Starting time: _____

**Time (min) from addition of amylase to color change[1]**

| Tube | pH | 1 | 2 | 3 | 4 | 5 | 6 | 7 | 8 | 9 | 10 | 11 | 12 | 13 | 14 | 15 |
|---|---|---|---|---|---|---|---|---|---|---|---|---|---|---|---|---|
| 1 | 5 | | | | | | | | | | | | | | | |
| 2 | 6 | | | | | | | | | | | | | | | |
| 3 | 7 | | | | | | | | | | | | | | | |
| 4 | 7[1] | | | | | | | | | | | | | | | |
| 5 | 8 | | | | | | | | | | | | | | | |
| 6 | 9 | | | | | | | | | | | | | | | |

[1] No enzyme present in Tube 4.

*Table 20-2. The effect of rennin on casein.*

| Tube | Reagents added to skimmed milk | Results[1] |
|------|-------------------------------|------------|
| 1 | Rennin | |
| 2 | Water | |

[1] Curds present or absent.

*Table 20-3. The effect of temperature and pH on hydrolysis of albumen by pepsin, determined by Biuret test for peptide*

| Tube | Reagents added to egg white | Temperature | Final color |
|------|-----------------------------|-------------|-------------|
| 1 | distilled water + pepsin | 37°C | |
| 2 | HCl + pepsin | 37°C | |
| 3 | NaOH + pepsin | 37°C | |
| 4 | HCl + pepsin | 0°C | |
| 5 | HCl + water | 37°C | |

*Effects of Digestive Secretions*

**Table 20-4.** *The digestion of macromolecules by pancreatin.*

**A. The effect of pH on the hydrolysis of starch by pancreatin, determined by $I_2$-KI test for starch. (+ indicates that a dark blue color was observed.)**

Time (min) from addition of pancreatin to color change[1]

| Tube | pH | 1 | 2 | 3 | 4 | 5 | 6 | 7 | 8 | 9 | 10 | 11 | 12 | 13 | 14 | 15 |
|------|----|---|---|---|---|---|---|---|---|---|----|----|----|----|----|----|
| 1 | 5 | | | | | | | | | | | | | | | |
| 2 | 6 | | | | | | | | | | | | | | | |
| 3 | 7 | | | | | | | | | | | | | | | |
| 4 | 7[1] | | | | | | | | | | | | | | | |
| 5 | 8 | | | | | | | | | | | | | | | |
| 6 | 9 | | | | | | | | | | | | | | | |

[1] No enzyme present in Tube 4.

*Table 20-4, continued*

## B. The effect of pancreatin on egg white, determined by Biuret test for peptide bonds.

| Tube | Reagents | Final color |
|------|----------|-------------|
| 1 | Egg white and pancreatin | |

## C. The effect of pancreatin and bile salts on the hydrolysis of fats, determined from pH changes indicated by litmus cream.

| Tube | Reagents | Final color |
|------|----------|-------------|
| 1 | cream + pancreatin | |
| 2 | cream + distilled water | |
| 3 | cream + pancreatin + bile salt | |
| 4 | cream + distilled water + bile salt | |

## Questions

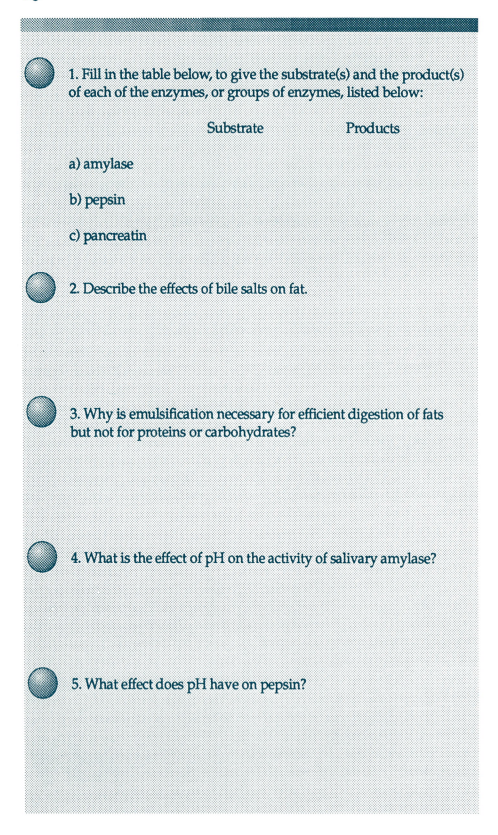

1. Fill in the table below, to give the substrate(s) and the product(s) of each of the enzymes, or groups of enzymes, listed below:

|  | Substrate | Products |
|---|---|---|
| a) amylase | | |
| b) pepsin | | |
| c) pancreatin | | |

2. Describe the effects of bile salts on fat.

3. Why is emulsification necessary for efficient digestion of fats but not for proteins or carbohydrates?

4. What is the effect of pH on the activity of salivary amylase?

5. What effect does pH have on pepsin?

6. What effect would fat digestion have on pH of the surrounding solution? _____ Explain.

7. Is there a difference between the optimum pH for salivary amylase and the optimum pH for pancreatic amylase? _____ Explain.

8. Describe the experimental controls in each part of this exercise:

a) Part 1 _____

b) Part 2 _____

c) Part 3 _____

d) Part 4 _____

9. What would happen to ingested fat if the bile duct were blocked or the gall bladder were removed?

Effects of Digestive Secretions

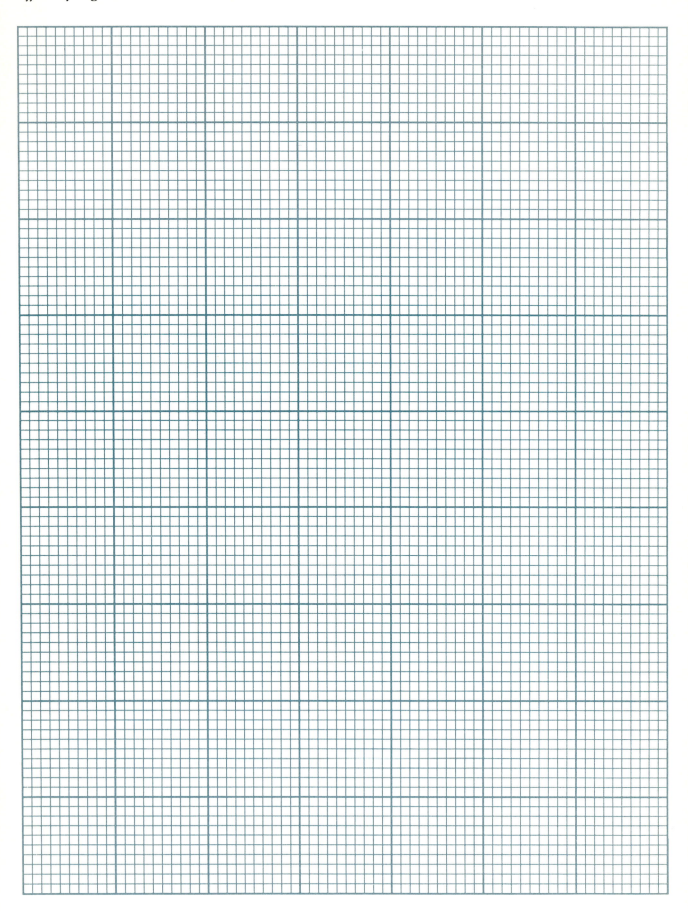

# Effects of Testosterone

*Reading assignment: text 309, 653-658, 660-661*

## Objectives

### Experimental

1. To measure the effects of testosterone injections on a) secondary sex characteristics b) weight gain, and c) gonad weight in male chicks.

### Conceptual

After completing this exercise and the reading assignment, you should be able to:

1. Describe the effects of testosterone on male chicks.

2. Describe the negative feedback mechanisms by means of which testosterone production is regulated.

## Background

This exercise will introduce you to some of the procedures used for experiments in endocrinology and will give you an opportunity to gain experience in experimenting on living animals over several weeks and performing clinical procedures, such as giving injections. You will administer hormone injections to male chicks and compare the injected chicks to control animals.

**Testosterone** is a male hormone, a **steroid** produced by the **interstitial cells** of the **testes**, that stimulates the production of sperm. (Note that the singular of testes is testis, just as hypotheses is the plural of hypothesis.)

Testosterone also stimulates the development of secondary sex characteristics. In human beings, the male secondary sex characteristics include a deep voice, facial hair, a narrow pelvis, increased bone mass and growth of the long bones, higher hematocrit, a higher ratio of muscle mass to total body weight in comparison to females, and, perhaps, sex-specific behaviors. Testosterone also produces male secondary sex characteristics in other species. In roosters, this hormone produces a large **comb** and **wattles** (Fig. 21-1), as well as **crowing** and **aggressive behavior**.

Because it produces an increase in body mass, testosterone is referred to as an **anabolic steroid**. You are probably aware that some athletes take

**Figure 21-1.**
*Comparison of a hen (left) and a rooster (right), showing enlarged comb and wattles of the rooster.*

anabolic steroids such as testosterone or testosterone derivatives to increase muscle mass (Fig. 21-2). Although the use of such drugs without a prescription is illegal and is grounds for disqualification in many competitions, the use of anabolic steroids among athletes is thought to be widespread. After doing this experiment, you will be able to evaluate the physiological risks of this practice.

Testosterone production is regulated by a negative feedback control system (Fig. 21-3). High blood levels of testosterone inhibit production of **gonadotropin-releasing hormone (GnRH)** by the **hypothalamus**. Because GnRH controls the secretion of two **gonadotropins, luteinizing hormone (LH)** and **follicle-stimulating hormone (FSH)**, by the **anterior pituitary**, elevated levels of testosterone indirectly inhibit LH and FSH secretion. Testosterone also causes the testes to produce **inhibin**, which directly inhibits secretion of LH by the anterior pituitary. In turn, low levels of gonadotropins and releasing hormone result in decreased production of testosterone.

In this experiment, you will evaluate the effects of injected testosterone (compared to injections of saline and sham injections) on the characteristics of male chicks listed below:

1. **Secondary sex characteristics,**
   a) development of the comb
   b) behavior
      aggressiveness
      crowing
2. **Weight gain**, and
3. **Feedback control of testosterone production.**

You will assess feedback control of testosterone production by looking at testis weight at the end of the experiment.

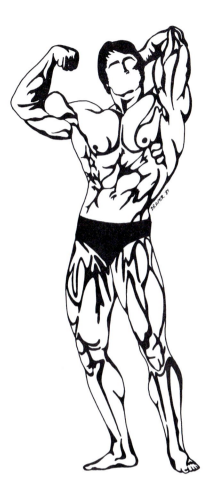

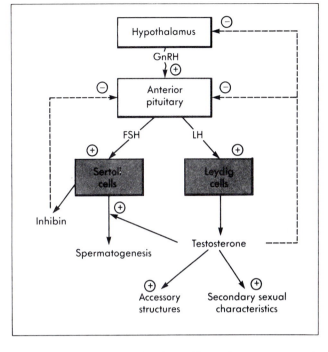

*Figure 21-2.
As methods of
training are
perfected,
athletes may be
tempted to use
anabolic steroids
to increase
muscle mass.*

**Figure 21-3.** *Negative feedback effects of testosterone and
inhibin from the testes regulate the anterior pituitary's
secretion of the gonadotropins LH and FSH.*

Before you begin this experiment, you should
have a mental model of how testosterone production is regulated. Summarize your model by stat-
ing one or more hypotheses.

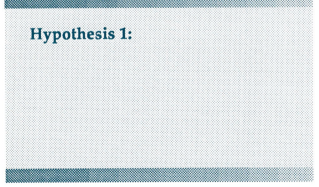

**Hypothesis 1:**

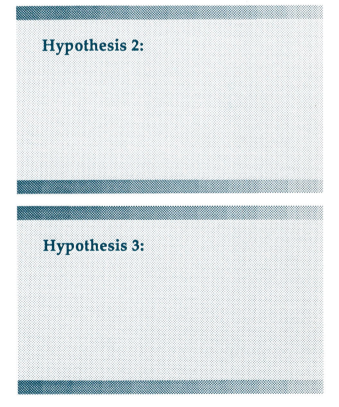

**Hypothesis 2:**

**Hypothesis 3:**

Next, specify your predictions about what will happen to the chicks treated with testosterone. In the list of predictions below, choose the best answer to fill in the blanks.

**Prediction 1:** Chicks treated with testosterone will display (an increase, a decrease, no change) _____ in **crowing behavior** in comparison to chicks that did not receive testosterone.

**Prediction 2:** Chicks treated with testosterone will display (an increase, a decrease, no change) _____ in **aggressive behavior** in comparison to chicks that did not receive testosterone.

**Prediction 3:** Chicks treated with testosterone will display (an increase, a decrease, no change) _____ in **comb weight** in comparison to chicks that did not receive testosterone.

**Prediction 4:** Chicks treated with testosterone will display (an increase, a decrease, no change) _____ in **body mass** in comparison to chicks that did not receive testosterone.

**Prediction 5:** Chicks treated with testosterone will display (an increase, a decrease, no change) _____ in **testis weight** mass in comparison to chicks that did not receive testosterone.

The students in each lab section will form five groups, and each group will study 3 male chicks. You will use three "treatments" (actually, one treatment plus two types of controls). One chick in each group will be assigned to each of the three treatments, as follows:

**Experimental (testosterone) group:** chicks will receive injections of testosterone in physiological saline.

**Saline control group:** chicks will be injected with a saline solution.

**Sham injection (handling control) group:** chicks will receive mock injections that will involve all the steps used for the other groups except that the syringe will only give an equivalent puncture wound.

Data will be obtained from the chicks during the initial lab period (Day 1) and every second day for the next 2 weeks. Each chick will be injected 7 times (on days 1, 3, 5, 7, 9, 11, and 13). At the end of the experiment, the chicks will be sacrificed so that you will be able to determine weights of the testes and combs.

## Procedure

Record your results in the laboratory report section at the end of this exercise.

The results from your entire laboratory section will be pooled. Because of this, and because the experiment extends through a period of 2 weeks, it is imperative that you thoroughly understand your assignment and responsibilities. You must be extremely meticulous in recording each item of procedure and all pertinent observations and data. The failure of a single student to perform an assignment correctly on a single occasion may jeopardize the experiment for the entire section. If you have any questions whatsoever concerning your assignment, be certain to consult your instructor.

### DAY 1

1. Band chicks. Your first assignment will be to band the chicks so that you can identify them in the future. Do not open the bands any wider than necessary to get them around the chick's leg, or they will be likely to fall off. Put one band on each leg (so you will still be able to identify the chick if one band comes off). Record the band numbers and colors on the Individual history sheet (Table 21-1). If at any time you notice that a band has come off, find it and put it back on or get a new one and record the change on the history sheet.

The three treatment groups will be color coded as follows:

Testosterone Group: green bands,
Saline Control Group: yellow bands,
Sham Injection Group: blue bands.

2. Obtain weights. Place each chick in a paper bag and weigh the chick plus bag. Record the weights on the individual history sheets (tables 21-1A, 21-1B, and 21-1C).

3. Inject chicks, using sterile technique as follows:

**Sterile technique.** Cleanse the syringe and syringe needle. Draw some 95% alcohol from the bottle marked "first rinse" into the syringe barrel and squirt it back into the bottle several times. Repeat the cleansing procedure in the second and third rinse solutions of alcohol. Place the empty syringe and syringe needle on a paper towel.

a. Testosterone Chick: Inject* the chick with 0.2 ml testosterone solution, as demonstrated by your instructor (0.2 ml solution contains 5 mg of testosterone dissolved in an aqueous solution of physiological saline). Record the preparation injected, the side injected, and any behavioral observations in Table 21-1A.

**\*Procedure for giving injections.** You will be giving subcutaneous (under the skin) injections. To do this, lift up the wing and swab the skin with a bit of cotton dipped in alcohol. Part the feathers and insert the needle through the loose skin but not into the muscle. Inject the solution. You should alternate between injecting the right and left sides.

The injection solutions and cleaning solutions will be kept at three different places in the lab to reduce the chance that bottles or syringes will get mixed up. You are responsible for being sure that you give each chick the proper treatment. If you discover that you have goofed, mark the inappropriate treatment on the appropriate history sheet (tables 21-1A, 21-1B, 21-1C), so that the results obtained from that chick can beproperly interpreted.

b. Saline Control Chick: Inject* the chick with 0.2 ml saline solution. Record the preparation injected, the side injected, and any behavioral observations in Table 21-1B.

c. Sham Injection Control Chick: Handle the chick as you did the other two chicks, but, instead of injecting a solution, just make a puncture wound. Air must not be injected. Be sure to use the same sterile technique that you used for the other groups. Record your observations in Table 21-1C.

## DAYS 3, 5, 7, 9, 11, and 13

Care of Chicks: Watering, feeding, and cleaning of each lab's chicks will be the responsibility of the students. The assignment of cleaning duties (which should be performed at least every other day) will be made in the first lab session for a 2-week period. You are all responsible for seeing that your chicks have enough food and, especially, enough clean water.

1. Repeat steps 2 to 4 from the procedure for Day 1. Record your results on the individual history sheets (tables 21-1A, 21-1B, 21-1C).

2. Describe comb growth, crowing behavior, and condition of each chick on the history sheets (tables 21-1A, 21-1B, 21-1C).

3. Replace any bands that have come off. Be sure to record the numbers of the new bands on the history sheets (tables 21-1A, 21-1B, 21-1C).

4. Check food and water of the chicks and replace if necessary. (Be certain that these are adequately supplied.)

5. Report any problems or discrepancies to your instructor immediately.

## DAY 15

This will be the regular meeting of your lab section 2 weeks after the day you began this experiment. Perform the following procedures for each chick:

1. Sacrifice the chick by compression over the pectoral muscles.

2. Weigh chick and record weight on the history sheet (tables 21-1A, 21-1B, 21-1C). Calculate cumulative weight gains (g) for days 3, 5, 7, 9, 11, 13, and 15 and enter these values on the history sheets (tables 21-1A. 21-1B, 21-1C).

3. Dissect out both testes. Give these to the instructor to weigh. Record testis weights on the history sheets (tables 21-1A, 21-1B, 21-1C).

4. Remove combs with a razor blade. Give combs to the instructor to weigh. Record comb weights on the history sheets (tables 21-1A, 21-1B, 21-1C).

5. Enter your results for each of the variables listed below on the summary sheets provided by your instructor.

> Initial body weight (g)
>
> Final body weight (g)
>
> Final weight (g)/Initial weight (g) (This will be a dimensionless ratio, because the grams in the numerator and the denominator will cancel out.)
>
> Testes weight (mg) (This should be the combined weight of both testes.)
>
> Testes weight (mg)/ Final body weight (g) (The dimensions for this ratio can be left as mg/g.)
>
> Comb weight (mg)
>
> Comb weight (mg)/Final body weight (g) (Again, your units will be mg/g.)
>
> Cumulative weight gain (g) for days 3, 5, 7, 9, 11, 13, and 15.

6. After your instructor has compiled summary sheets for all groups (this may take several days, if laboratory sections that meet on different days are contributing data) obtain copies of the completed summary sheets for each treatment group (testosterone, saline control, and sham injection). Use these data to calculate mean values for the variables listed in Step 5 for chicks in each of the three treatment groups. Enter these values in the appropriate places on the master data sheets (tables 21-2 and 21-3).

# Laboratory Report

### Exercise 21: Effects of Testosterone

Name: _____

Date: _____

Lab Section: _____

## Analyzing Your Data

Prepare a report of your findings. Refer to your hypotheses and predictions at the beginning of this experiment. Were your predictions fulfilled?

If not, was this because of something you failed to take into consideration at the beginning of the experiment? Or, was it because of problems with the way the experiment was carried out? Did you learn something?

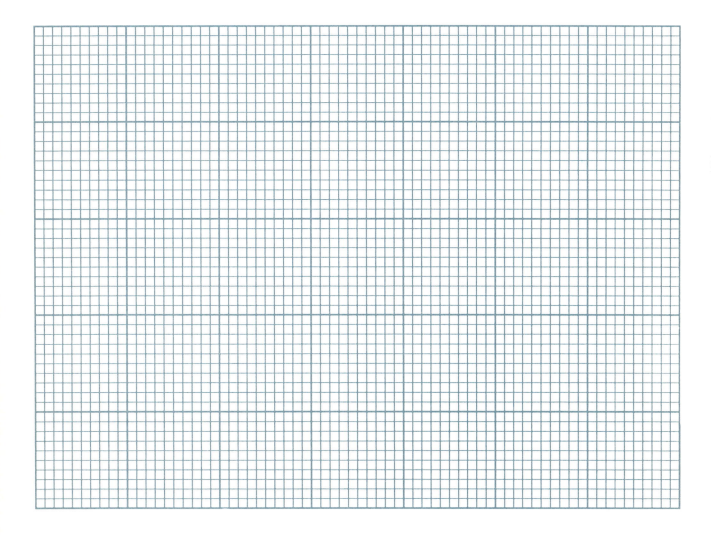

You may wish to include one or more graphs in your report. For instance, you may find it useful to graph changes in one or more of the experimental variables during the course of the experiment.

Before you begin writing your report decide which relationships you want to show graphically. Make sure that all graphs include a title and that all lines on each graph are clearly labeled.

*Table 21-1A. Effect of testosterone on male chicks: Individual history sheet.*

| Lab Section: | Group: | Band number: | Color: |
|---|---|---|---|

**Weight of comb on Day 15:**　　　　**Weight of testes on Day 15:**

| Day | Date | Time | Weight (g) | Cumulative weight gain (g)[1] | Preparation injected | Amount injected | Side injected | Observations[2] |
|---|---|---|---|---|---|---|---|---|
| 1 | | | | — | | | | |
| 3 | | | | | | | | |
| 5 | | | | | | | | |
| 7 | | | | | | | | |
| 9 | | | | | | | | |
| 11 | | | | | | | | |
| 13 | | | | | | | | |
| 15 | | | | | | — | — | — |

[1] Cumulative weight gain for $\text{day}_n$ = (weight on $\text{day}_n$) - (weight on $\text{day}_1$).
[2] Appearance of comb, crowing behavior, general physical condition.

**Table 21-1B.** *Effect of saline on male chicks: Individual history sheet.*

| Lab Section: | Group: | Band number : | Color : |
|---|---|---|---|

| Weight of comb on Day 15: | Weight of testes on Day 15: |
|---|---|

| Day | Date | Time | Weight (g) | Cumulative weight gain (g)[1] | Preparation injected | Amount injected | Side injected | Observations[2] |
|---|---|---|---|---|---|---|---|---|
| 1 | | | | — | | | | |
| 3 | | | | | | | | |
| 5 | | | | | | | | |
| 7 | | | | | | | | |
| 9 | | | | | | | | |
| 11 | | | | | | | | |
| 13 | | | | | | | | |
| 15 | | | | | — | — | — | |

[1] Cumulative weight gain for $day_n$ = (weight on $day_n$) - (weight on $day_1$).
[2] Appearance of comb, crowing behavior, general physical condition.

*Effects of Testosterone*

**Table 21-1C.** *Effect of sham injections on male chicks: Individual history sheet.*

| Lab Section: | Group: | Band number: | Color: |
|---|---|---|---|

**Weight of comb on Day 15:**    **Weight of testes on Day 15:**

| Day | Date | Time | Weight (g) | Cumulative weight gain (g)[1] | Preparation injected | Amount injected | Side injected | Observations[2] |
|---|---|---|---|---|---|---|---|---|
| 1 | | | | — | | | | |
| 3 | | | | | | | | |
| 5 | | | | | | | | |
| 7 | | | | | | | | |
| 9 | | | | | | | | |
| 11 | | | | | | | | |
| 13 | | | | | | | | |
| 15 | | | | | — | — | — | |

[1] Cumulative weight gain for $day_n$ = (weight on $day_n$) - (weight on $day_1$).
[2] Appearance of comb, crowing behavior, general physical condition.

*Table 21-2. Master data sheet: mean values for body weight, comb weight, and weight of testes: mean values for pooled data from all sections.*

| | Body Weight (g) | | | Testes | | Comb | |
|---|---|---|---|---|---|---|---|
| Treatment | Initial | Final | Initial/Final | Weight (mg) | mg/g[1] | Weight (mg) | mg/g[1] |
| Testosterone | | | | | | | |
| Saline injection | | | | | | | |
| Sham injection | | | | | | | |

[1] Weight divided by body weight in g.

*Table 21-3. Master data sheet for body weight gain: mean values for pooled data from all laboratory sections.*

| | Body weight gain (g) | | | | | | |
|---|---|---|---|---|---|---|---|
| | Day | | | | | | |
| Treatment | 3 | 5 | 7 | 9 | 11 | 13 | 15 |
| Testosterone | | | | | | | |
| Saline injection | | | | | | | |
| Sham injection | | | | | | | |

# Questions

1. Describe how the animals treated with testosterone differed from the control animals in each of the characteristics listed below.

a) growth rate      _____

b) behavior      _____

c) gonad weight      _____

d) weight of secondary
sex structures      _____

2. What caused the differences between the animals treated with testosterone and the control animals?

3. List the hormones involved in the negative feedback loops that regulate testosterone production and the functions of and organs that produce each hormone.

4. Why might testosterone administered to normal males in the hope of improving athletic performance also cause sterility?

5. If testosterone is an anabolic hormone and causes the growth spurt of puberty, why might chicks treated with testosterone weigh less than controls?

6. Explain the purpose of a) the saline control and b) the sham injection control in this experiment.

7. What is the advantage of calculating the following ratios: final weight/initial weight, testes weight/body weight, and comb weight/body weight, instead of just looking at final weight, testes weight, and comb weight?

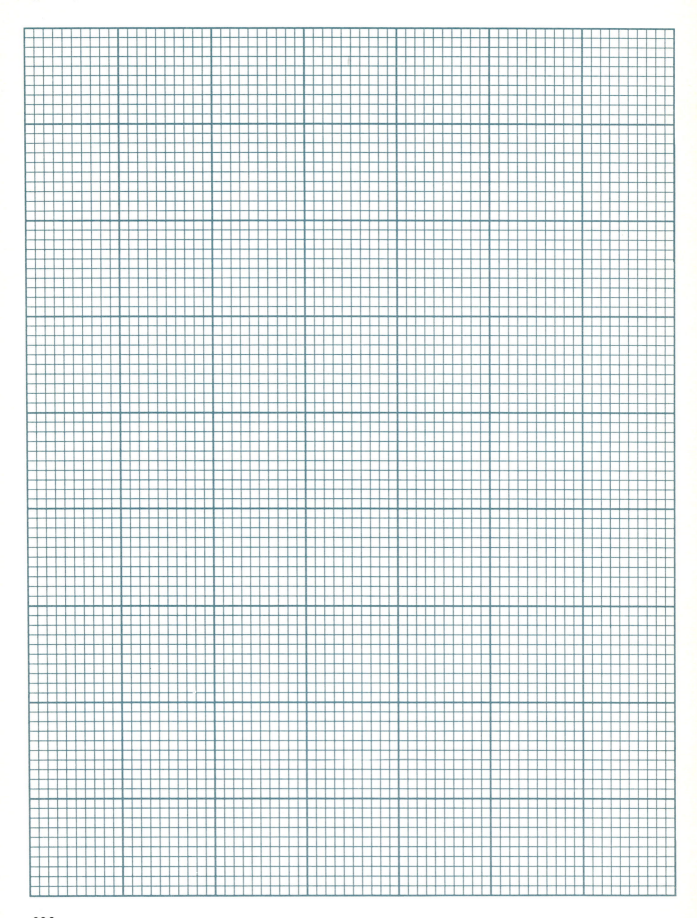

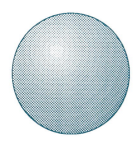

# A Computer Simulation of Muscle Contraction
## Mechanical Properties of Active Muscle

*Mechanical Properties of Active Muscle*
*by Richard A. Meiss,*
*Indiana University School of Medicine*
*is available from*

COMPress/Division of Queue, Inc.
338 Commerce Drive
Fairfield, CT 06430
800-232-2224, 203-335-0908

This program allows you to stimulate a muscle with properties typical of frog gastrocnemius. It actually consists of several separate programs, each specialized for a particular type of experiment.

In each case, you will be able to vary the stimulus intensity and to measure the resulting force or isometric tension.

## Before you start:

1. Be familiar with the following terms: **threshold, twitch, tetanus, isometric, isotonic, summation**.

2. Have ready a pad of ruled paper for collecting data and keeping notes and a pad of tracing paper to record data from the computer screen. These aids are particularly important if you are to prepare a report on this exercise.

## Getting Started:

If necessary, boot the computer using the system disk. Then replace the system disk with the program disk and type

 **GO [RETURN].**

Pass through the introductory screens until you reach the main menu, headed "Index of Programs".

### I. Isometric Twitch

At the main menu, press

 **2** to enter the program titled "Electrical Stimulation".

When the menu for this program comes on the screen, press

 **3** to enter the subprogram titled "Single Stimulus".

You will see a screen containing a graph of force versus time and a stimulator. The time scale runs from zero (the point at which the stimulus is delivered) up to 1 sec. The time your model muscle needs to complete a twitch will vary depending on the speed of your computer, but in any case it will take longer than 1 second of real time. One advantage of the computer simulation is that time is slowed down, so you can observe features of the twitch that could not be captured by the chart recorders used in most student laboratories in an experiment with real muscle.

The intensity of stimulation is set using the + and – keys as indicated. Set the stimulus intensity to

⦿ **1 V** and press

⦿ **S**

to deliver a stimulus. Is any force developed as a result? Increase the stimulus voltage to

⦿ **2 V**

and stimulate again. Notice that each trace is saved on the screen.

⦿ Continue increasing the intensity by steps of **1 V** and then stimulating, until you reach **10V**.

Notice that the maximum force (the peak of the twitch) increases rapidly at first and then moreslowly until a voltage is reached at which furtherincreases in voltage do not increase the maximum force.

⦿ Fasten a piece of tracing paper to the screen and trace the family of force curves, including the horizontal and vertical scales and labeling each curve with the corresponding stimulus intensity. When you have done this, press

⦿ **E** to erase the traces.

⦿ Now go back and determine the threshold more precisely by varying the voltage in **0.1V** steps over the interval in which force began to be developed. The threshold you determined by this method is _____V.

## II. Summation of Contractions

Return to the menu headed "Electrical Stimulation of Skeletal Muscle" by typing

⦿ **Q.**

Choose option **4**, "Multiple Stimulus". The screen looks much like that of Part I, except that you can set the rate of stimulation as well as the intensity. Referring to your data from Part I, set the stimulus

voltage to a value that will give maximum force. The default rate is 1/sec. Press

⦿ **K** to trigger a stimulus.

You should get a single twitch. How long does the twitch last? If the rate were 3/sec, successive stimuli would occur before the initial contraction was over, making summation of force possible.

⦿ Predict what the maximum force generated will be if the stimulus rate is increased to 3/sec. Enter your prediction here: _____.

Now press

⦿ **R**

to call up a box that allows you to increase the rate. Set the rate to

⦿ **3/sec**.

Exit the box by pressing

⦿ **V.** Press

⦿ **T** to limit the stimulation to 1 sec. (1 screen worth of time). Press

⦿ **K** to trigger stimulation. How closely did the result match your prediction?

How long does it take the muscle to reach peak force after it is stimulated with a single shock? Using this information, can you predict the stimulus rate at which the muscle will contract smoothly in a tetanus? Enter your prediction here: _____.

⦿ Continue to increase the stimulus rate and store single traces on the screen. How closely does the result match your prediction? (You may not have been able to test your prediction very precisely because the number of rate options the program offers is limited). At this point the screen should show a graph record with 5 traces, one corresponding to each stimulus rate you tried.

Make a tracing of this record, including the horizontal and vertical scales and labeling each trace with the corresponding stimulus rate.

## III. *Load and Shortening in Afterloaded Twitches*

Return to the main menu and choose option **4**, Isotonic Contractions.

In this program the muscle performs afterloaded isotonic contractions and you can vary the weight of the load and the position of the load support. The upper graph shows the muscle tension; the lower graph muscle length. Set the length at **6**. Set the weight at **1**. Press

    **S** to stimulate.

Notice that the resulting contraction can be divided into three phases. In the first, muscle tension is rising from its initial value to a value equal to the load. During this phase, the sarcomeres are shortening but the series elastic elements are stretching; the contraction is isometric during this phase as indicated by the box on the screen. In the second phase, the muscle shortens and lifts the load. The contraction is isotonic during this phase. The last phase begins when tension in the muscle falls to a value equal to that of the load and continues until the muscle has completed its relaxation. As in the first phase, the third phase is isometric.

Increase the load to **2** and stimulate again. Notice two changes:

> 1. the time spent in the two isometric states is longer, at the expense of the time spent in shortening.

> 2. when the muscle does begin to shorten, it lifts the load more slowly than when it was more lightly loaded.

The effect of these two changes is to reduce the distance that the load is lifted.

Increase the load to **3**. Can the muscle lift the load at all? Now increase the length to **9**, keeping the load the same. This length change is essentially equivalent to preloading the muscle.

    Restimulate.

Can the muscle now lift the load?

## IV. *Effect of Load on Muscle Work*

The total work done by the muscle in an isotonic twitch is equal to the weight of the load (typically measured in g for a muscle of this size) multiplied by the distance the weight is lifted (typically measured in cm). The corresponding units of muscle work are g-cm. If you have carried out the exercise on living muscle (Exercise 9), you will recall that the work done in a single twitch is zero if the load is zero. It rises to a maximum as the load is increased and then falls back to zero as the load is made so heavy that it cannot be lifted by the muscle. The power output of the muscle, or rate of doing work, is equal to the rate of shortening multiplied by the load. The corresponding units of power output are g-cm/sec. The power output curve is similar in shape to the curve of work versus load. There is thus an optimum load and shortening velocity at which muscle power output is maximum.

Return to the main menu and choose option **7**, The Biker. The idea of this exercise is to keep the biker's power output maximal by adjusting the shortening velocity (changing gears) as the load is changed by wind. If you understand the concept of muscle power output, this program is self-explanatory and fun.

# The Cardiovascular System and the Baroreceptor Reflex
## Circsyst 2.0

*Circsyst 2.0 is available from*

*Joseph Boyle III, M.D.*
*Department of Physiology*
*New Jersey Medical School*
*100 Bergen Street*
*Newark, N.J. 07103*

This program allows you to manipulate some properties and variables of the cardiovascular system and see the effects on other variables and properties of the system. The program contains explanatory screens, so an extensive introduction is not necessary.

The cardiovascular system is characterized by complex interactions of multiple variables, not all of which are fully understood. This program is a considerable simplification of the system. The values it returns for given inputs will not always correspond perfectly to those that would usually be obtained from a normal live animal or patient.

## Getting Started:

Unless directed otherwise by your instructor, boot the program from the disk by logging onto the drive containing the program disk and typing

 **AUTOEXEC.**

After the program has loaded, you must follow the instructions given on the first screen to set the speed.

After some screens of introductory material, the program offers you the choice of having data displayed as either

1. volume and pressure versus time
2. left ventricular pressure-volume loops

Option **1** is most useful for the simulations outlined in this exercise.

The program then offers you a choice of topics to study. Choose option **3** (Individual Parameters of Cardiovascular Performance) to run simulations. Options **1** and **2** give background information that may be useful for review. Choosing option **3** gives a screen with a list of the variables you can alter. This will be referred to as the Variables Menu. You will be directed to change one or more of these in subsequent parts of this exercise.

## The Traces Screen

After you have entered any changes in the Variables Menu, pressing

 **RETURN**

starts the simulation. The screen shows traces of aortic pressure and left ventricular pressure such as would be recorded from pressure transducers

342

inserted into these sites. It also shows a trace of ventricular volume.

After a short period of control record, the program begins to respond to the values you entered. The traces continue until you press

 **S** to stop the simulation.

If you don't stop the simulation before the traces reach the right-hand side of the screen, the traces will wrap around and continue running, but the program does not display an account of elapsed time after the first wraparound.

After you stop the simulation, pressing

 **RETURN**

gives you a screen that summarizes the important variables of the system as they existed during the control period and at the instant you stopped the simulation. This will be referred to as the Data Summary Screen. You can get printouts of the summary screens if your computer is attached to a printer and has a Print Screen command. Print Screen will not give you a faithful copy of the traces screen. Making a copy of the data summary screens with a printer or by hand is the most practical way to keep a record of your experiments. In some cases it will be necessary to use your copies of summary screens to compare results from different simulation runs.

After you have viewed the Data Summary screen, you are offered an opportunity to alter the display format or move to another section of the program. If you are going to run another simulation, answer **N**(o) to return to the Variables Menu.

## I. Effects of Changing System Properties and Variables in Absence of Reflexive Compensation

### A. Heart Rate

At the variables menu, press

 **4** to change the heart rate.

Enter **120** to cause a moderate tachycardia. Press

 **RETURN** to start the simulation. Answer

 **N**

when the program asks if the baroreceptor reflex is to be allowed to operate. Let the traces continue until a new stable arterial pressure is reached, then press

 **S** to stop the simulation. Press

 **RETURN**

to see a summary of the status of the system at this point in time. Note the following changes:

Aortic pressure is increased.
Cardiac output is increased.
LVESV is decreased.
Both LVEDV and LVESV are decreased.
The stroke volume (SV) is equal to LVEDV - LVESV.

 Calculate the SV before and after the increase in heart rate. Did it change much?

You should be able to predict and explain each of the changes shown in the data screen. You may wish to explore the effects of several different heart rates before leaving this part of the exercise.

### B. The Effect of Changing Blood Volume

From the Variables Menu, enter **5** to change blood volume. Enter an infusion rate of **-10**. This will be equivalent to severe bleeding. Note that the bleeding will continue until you enter

 **S** to stop the simulation or until the imaginary patient dies of hypovolemic shock. Choose

 **no baroreceptor reflex**.

 Run the simulation until the traces reach the right side of the screen and then press

 **S**. Press

*Circsyst 2.0*

 **RETURN**

to see a data summary. Note the effects of bleeding on aortic pressure, cardiac output and the other system variables.

Now restart the simulation at the same infusion rate and allow it to run until the patient dies. Make a note of the time that elapsed from the start of bleeding to death. You should be able to predict the effect on system variables of increasing the blood volume.

Rerun the simulation with a positive infusion rate to test your predictions.

### C. The Effect of Changing Peripheral Resistance

Return to the variables menu and choose 6 (arteriolar resistance). Enter

 **0.5** to cause arteriolar dilation.

Will this change increase or decrease the total peripheral resistance of the vascular system? Run the simulation without the baroreceptor reflex. When the values stabilize, press

 **S** and then

 **RETURN**

to call up the data summary screen. Note that while there is a decrease in mean aortic pressure, CO increases. Calculate the change in stroke volume. What caused stroke volume to increase?

### D. The Effect of Changing Ventricular Contractility.

Myocardial contractility is a system property. Changes in contractility can be caused by drugs and neurotransmitters; such effects are termed **inotrophic effects**. Return to the Variables Menu and choose 3 (ventricular contractility). Enter a value of **2**. In the living system, such a positive inotrophic effect could be obtained with

epinephrine or with drugs (beta-adrenergic agonists) that activate the beta receptors of the myocardium.

⬤ Run the simulation without reflexive compensation. Stop it after new stable values are reached.

Note the increases in mean aortic pressure, stroke volume and cardiac output. Why does LVEDV decrease? Why does LVEDP decrease?

### II. The Baroreceptor Reflex

The baroreceptor reflex is one of a number of neural and hormonal control systems that regulate arterial blood pressure, blood volume and plasma composition. The regulated variable for the baroreceptor reflex is the mean arterial pressure (MAP) as measured by the baroreceptors of the carotid sinus. The MAP is a time-averaged value and is not equal to the simple arithmetical mean of the systolic and diastolic pressures, but for purposes of approximation you can take the mean pressure to be about equal to the diastolic pressure plus 1/2 of the pulse pressure.

You can examine the effect of the baroreceptor reflex by running four simulations that perturb the MAP and trigger a reflexive compensation.

### A. Bleeding

Choose option **5** (blood volume) from the Variables Menu. Set the infusion rate to **–10** (severe bleeding). This is the same blood loss rate you used in Part IB except that in this case you will choose to leave the reflex intact.

⬤ Stop the simulation after 20 min and examine the data summary.

Note that heart rate is increased and cardiac output decreased, but MAP is almost unchanged. Blood volume has dropped by almost half. What changes has the reflex made in the properties of the cardiovascular system to keep blood pressure constant in the face of such a large change in blood volume?

🔵 Restart the simulation with the same blood loss rate and let it run until the imaginary patient dies.

Compare the time from the beginning of bleeding to death in the presence of the reflex to that that you noted in Part IB. What could account for the difference? Do you think the program is simulating arterial bleeding or venous bleeding?

## B. Injection of Epinephrine

Recall that epinephrine has several effects on the cardiovascular system: its cardiac effects are increases in heart rate and ventricular contractility. Its effect on peripheral resistance are complex - low doses primarily affect beta adrenergic receptors, causing mainly vasodilation and a drop in resistance. At larger doses the alpha adrenergic effect predominates and total peripheral resistance increases. We will model a dose that increases peripheral resistance.

Enter the following changes: Ventricular Contractility: **2**; Heart Rate: **120**; Arteriolar resistance: **1.5**. Run the simulation first with the reflex disabled.

Note that MAP almost doubles by 20 min and continues to remain high as long as you let the simulation run.

🔵 Repeat the simulation with the same values and this time leave the reflex intact.

Notice the profound drop in heart rate. MAP changes only slightly, as expected. Can you explain the large increase in stroke volume?

## C. Effects of Aerobic Training on Cardiovascular Performance at Rest

Regular exercise causes an adaptive increase in ventricular muscle mass. Forms of exercise that cause sustained increases in heart rate and oxygen consumption are sometimes called

**aerobic training** and are particularly effective in increasing cardiac performance. You can model the effect of such training on the heart by increasing ventricular muscle mass to 190. This change takes place instantly in the simulation, but requires weeks or months of conditioning in reality. How would you expect aerobic training to affect each of the following at rest: heart rate, cardiac output, mean arterial pressure (MAP)?

Reset ventricular muscle mass to **190** and

🔵 run the simulation with the baroreceptor reflex intact.

How do the results compare with your predictions for resting values?

## D. Heart Failure

Heart failure may be simulated by decreasing the ventricular contractility. Set the ventricular contractility to **.5** and

🔵 run the simulation with the reflex disabled.

Note that MAP drops severely and CO is almost halved, while CVP rises. In the absence of the baroreceptor reflex, the only feedback control loop in the system is the Frank-Starling law of the heart. As the failure takes effect, the central venous pressure increases, stretching the ventricle during diastole until the CO and MAP stabilize at new lower values.

🔵 Repeat the simulation with the reflex intact.

How well does the reflex compensate for the decrease in contractility? How do the changes in CO and MAP with the reflex intact compare to those without the reflex?

# Resting and Action Potentials in a Single Axon
## The SPIKE Program

*The SPIKE Program is available from*

*John Cornell*
*3032 Maine Prairie Road*
*St. Cloud, MN 56301*

The axons of vertebrate neurons are typically less than 100 microns in diameter. It is usually not possible for students to record the activity of single axons of this size, but compound action potentials representing the sum of activity of many axons in a nerve can readily be recorded using extracellular electrodes (see Exercise 7). "Giant" axons (up to 1 mm in diameter) are found in some invertebrates, such as the squid, cockroach and earthworm. Even in these axons, however, it is not possible to measure the transmembrane voltage using equipment typically available to students. This program offers you the chance to experiment with a single nerve axon with properties similar to those of giant axons. The program lets you set the stimulus intensity and duration and control the extracellular $Na^+$ concentration.

## Before you start the program:

1. Review the following terms, concepts and equations: **resting potential, $Na^+$ gradient, $K^+$ gradient, $Na^+$ equilibrium potential, $K^+$ equilibrium potential, threshold, absolute refractory period, relative refractory period, $Na^+$ gates, $K^+$ gates, $Na^+$ inactivation, Nernst equation, Goldman Hodgkin Katz equation**

## Getting Started:

### I. Effect of Transmembrane $Na^+$ Gradient on Resting Potential

Start the SPIKE program by inserting the program disk and typing

 **SPIKE**.

The program asks for your student ID number. Then it gives you two pieces of information that you must write down: the intracellular $Na^+$ concentration and the temperature. This combination of values will be different for different users of the program. When you have recorded these values, press

 **ENTER**.

Next, the program asks you to enter a value for extracellular $Na^+$ concentration. The normal human extracellular $Na^+$ concentration is 136-142 mEq/L. Enter a value of **140 mEq/L**. This concentration will be used as the "normal" value. Press

 **ENTER**

to advance to the working screen of the program. You will see graph axes of membrane potential versus time. The horizontal dashed line crossing

346

| Extracellular Na$^+$ concentration (mEq/L) | Na$^+$ gradient (mEq/L) | Resting potential (mV) |
|---|---|---|
| 140 | | |
| 80 | | |
| | | |
| | | |
| | | |
| | | |
| | | |
| | | |

the graph represents the resting potential of your axon. According to the Goldman equation, the exact value of the resting potential is determined by the temperature, the transmembrane gradients of Na$^+$ and K$^+$, and the resting permeabilities of the axon to these ions. In this program all of these values except the extracellular Na$^+$ concentration are preset and cannot be changed. You determined the current Na$^+$ gradient when you chose an extracellular Na$^+$ concentration of 140 mEq/L.

⬤ Calculate the value of the Na$^+$ gradient by subtracting 140 mEq/L from the intracellular Na$^+$ concentration you recorded when you started the program, and enter it in the table above. Then enter the resting potential shown on the screen now. Now press

⬤ N to change the extracellular Na$^+$ concentration.

Try **80 mEq/L**. What effect did this have on the resting potential?

Compute the new transmembrane Na$^+$ gradient and enter it below, along with the resting potential. Continue in this way until you have a record of the resting potential and corresponding Na$^+$ gradient at at least ten different extracellular Na$^+$ concentrations.

Plot the data on the graph provided on page 348. Note that the horizontal axis has a logarithmic scale. Can you explain why this is appropriate?

## II. The Threshold

Now you are ready to try stimulating the axon. Set the extracellular Na$^+$ concentration to the normal value – **140 mEq/L**.

Note that the units of stimulus intensity are microA/cm$^2$, the amount of current that flows through 1 cm$^2$ of membrane surface, a direct measure of stimulus intensity. If you have used an electrical stimulator to stimulate living tissue, you may remember that the voltage of the shock

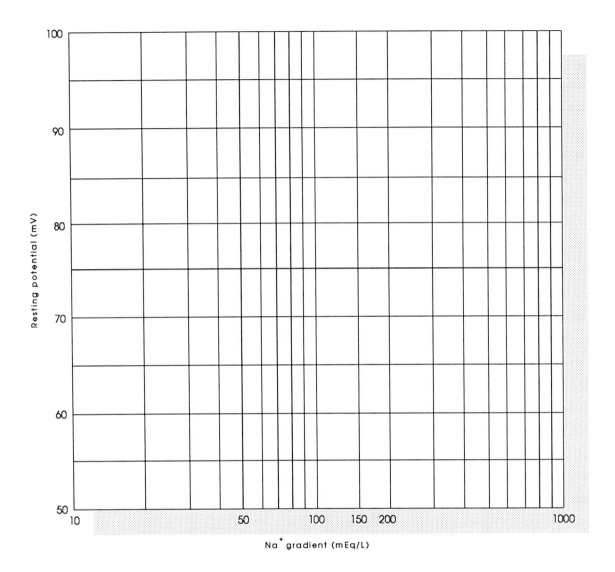

could be preset. The intensity of the resulting current through the stimulated tissue depended on the spacing between the probe electrodes, the size and shape of the stimulated tissue, and other factors.

Set the stimulus intensity to **20 μA/cm²** and the duration to **1 msec**.

 Deliver a stimulus.

Allow the trace to reach the right side of the screen, then stop it by pressing

 **THE SPACEBAR**

What happened? Is the resulting deflection an action potential? Now press

 **TAB**

to reset the simulus parameters without losing the trace you just recorded. Reset the stimulus intensity to **5 microA/cm²** and try again.

Notice that the voltage deflection is larger, but that the membrane potential still returns to the resting value promptly after the stimulus pulse is over. Continue to increase the stimulus intensity until an action potential results. With small alterations of the stimulus intensity, you can determine the

| Extracellular Na$^+$ concentration (mEq/L) | Threshold stimulus intensity (microA/cm$^2$) | Peak voltage |
|---|---|---|
| 60 | | |
| 140 | | |
| 200 | | |

threshold very precisely. Enter the value in the table above.

The resting potential is determined by the relative permeability of the axonal membrane to Na$^+$ and K$^+$ and the magnitudes of the concentration gradients for the two ions. As you found in the first part of this exercise, increases in the extracellular Na$^+$ concentration (which cause increases in the Na$^+$ concentration gradient) cause depolarization. Increases in the K$^+$ gradient would have the opposite effect (hyperpolarization). With the usual concentration gradients, the resting potential is near the K$^+$ equilibrium potential (–90 mV to –110 mV) and far from the Na$^+$ equilibrium potential (about +60 mV) because both the K$^+$ concentration gradient and the membrane's permeability to K$^+$ are greater than the corresponding values for Na$^+$. The high resting permeability to K$^+$ is due to the presence of "leak" channels for K$^+$ which remain open at rest.

When the axon membrane is at its resting potential, the inward current of Na$^+$ and the outward current of K$^+$ are equal. When the stimulus is delivered, two competing processes occur. Since the membrane potential is driven farther away from the K$^+$ equilibrium potential, the outward K$^+$ current increases. At the same time, the depolarization causes some voltage-sensitive Na$^+$ channels to open. The number of open channels increases with stimulus intensity and stimulus duration. If enough channels open, the inward current exceeds the outward current, and a positive

feedback cycle starts, with more depolarization causing more channels to open. An action potential results.

The threshold stimulus intensity is determined by the concentration gradients of Na$^+$ and K$^+$ and the number of Na$^+$ channels that can be opened relative to the number of K$^+$ channels open. Thus the threshold is not an unvarying value - it differs for different neurons and for the same neuron under different conditions. What effect would decreasing the Na$^+$ gradient have on the threshold intensity? How about increasing the gradient? What effect would increasing or decreasing the stimulus duration have on the threshold intensity?

Now test your predictions. If necessary, press

 **ESCAPE**

to clear the screen. Let the stimulus duration remain at 1 msec. Change the extracellular Na$^+$ concentration to **60 mEq/L** and determine the threshold, keeping the stimulus duration at 1 msec. What differences do you see in the form of the action potentials in the two Na$^+$ concentrations? Why is the peak height of the action potential higher in the higher Na$^+$ concentration? Now change the extracellular Na$^+$ concentration to **200 mEq/L** and redetermine the threshold. Enter the values the table on page 350. How does the peak height in 200 mEq/L Na$^+$ compare to that in 150 mEq/L Na$^+$?

| Stimulus duration (msec) | Threshold intensity (microA/cm²) |
|---|---|
| 0.1 | |
| 0.5 | |
| 1.0 | |
| 10.0 | |

Return the extracellular Na⁺ concentration to **140 mEq/L**. Conduct an experiment in which the stimulus duration is the variable. Use the durations given in the table on 350 and enter the corresponding threshold intensities.

## III. *The Refractory Period*

In this part of the exercise you will examine the effect of delivering a second stimulus at various time intervals after an action potential has been initiated. To deliver paired stimuli, you must press

 **THE SPACEBAR**

to stop the trace at the desired time for the second stimulus. Then reset the stimulus intensity and press

 **ENTER**

to deliver the second stimulus. Do not press **TAB** or **ESCAPE** between the first and second stimuli. Getting the timing of the second stimulus right may require a little practice, and there will be small variations in the timing, depending on how your brain's speed compares to that of your computer.

The intensity of the first stimulus should always be **100 microA/cm²**. The duration of all pulses should be 1 msec. For your first experiment, stop the trace at **3 msec**. Then restimulate.

Does a second action potential result? Press

 **ESCAPE**

to clear the screen and repeat, increasing the intensity of the second stimulus (200 microA/cm² is the maximum permissible value). Why is the axon still absolutely refractory at 3 msec, even though the upstroke of the action potential is already over?

Clear the screen again and try giving the second stimulus at **5 msec**. If you can get a second action potential at this time, determine the threshold stimulus intensity necessary. Plot it on the graph on page 351.

Notice the difference in spike heights between the two action potentials. Why does this difference occur?

Follow the same procedure to determine the threshold for a second action potential at 10 msec and at 15 msec. About how long after the first stimulus does the absolute refractory period last? About how long does the relative refractory period last? Why is the threshold so high at the beginning of the relative refractory period? What change causes the threshold to return to its usual value at the end of the relative refractory period?

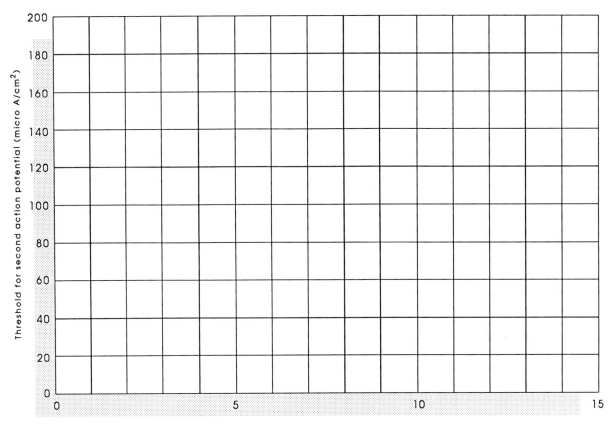

Threshold for second action potential (micro A/cm²) vs. Time interval between first and second stimuli (msec)

## IV. Repetitive Action Potentials and the Neural Code

So far, you have shown that, if the stimulus is intense, quite brief stimuli may be adequate to trigger an action potential. You also showed that fairly weak stimuli may trigger an action potentital if they continue for several msec. What happens if the stimulus duration is even longer? Stimuli that last for several tens of msec or longer can cause repetitive action potentials. The ability of excitable cells to respond to steady stimulation with a series of action potentials is particularly important for sensory receptors. Most sensory stimuli last for at least several tens of msec. Sensory neurons encode the intensity of stimuli in the frequency of their action potentials. The behavior of the excitable membrane of the input segment of the receptor provides the basis for frequency-intensity

coding. Using the model axon, you can simulate the behavior of a sensory receptor that is receiving a steady or **tonic** stimulus.

Reset the extracellular Na$^+$ concentration to **140** if necessary. Set the stimulus duration to **20 msec**. At some stimulus intensity your axon should give more than one action potential in response to this long depolarization. Find the lowest intensity that gives more than one action potential and measure the time that elapses between the peaks of the first and second action potentials (the interspike interval). Enter the data in the table on page 352.

Now increase the stimulus intensity. An increase of only a few microA/cm$^2$ should significantly shorten the interspike interval. Continue increasing the stimulus intensity and measuring interspike interval. After the intensity enters the range

| Stimulus intensity (microA/cm²) | Interspike interval (msec) | Spike frequency (spikes/sec) |
| --- | --- | --- |
| | | |
| | | |
| | | |
| | | |
| | | |
| | | |
| | | |
| | | |
| | | |
| | | |
| | | |

of several tens of microA/cm², you will need to make increasingly larger changes in intensity to elicit measurable changes in interspike interval. Obtain at least 10 data points that cover the range between the minimum intensity and 200 microA/cm².

The interspike intervals, which have units of msec, can be converted to frequency values (in units of spikes/msec) by dividing each interspike interval into 1000.

Convert intervals to frequencies in the table above and plot your data on the axes provided in the graph on page 353.

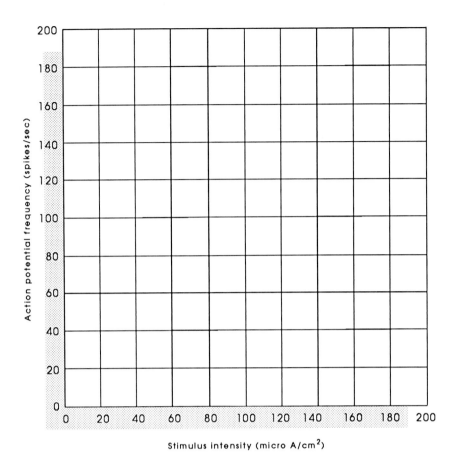

# Appendix 1– Interpreting and Preparing Graphs

Graphs are a useful way to convey a great deal of information, if you understand how to interpret them. Graphs provide pictorial representations of relationships between variables. The saying, "One picture is worth a thousand words," is true in this case. Often it is much easier to understand what is going on if you look at a graph of some data than it is if you read a description of the relationships or see the same information presented in a table.

Usually when we graph the relationship between two variables, we are interested in knowing how one (the **independent** variable) influences the other (the **dependent** variable). We graph values for the independent variable on the horizontal or *x* **axis**, also called the **abscissa**. (Note that the plural of axis is **axes**, the same as the plural of axe, but pronounced ax-eez.) Values for the dependent variable are graphed on the vertical or *y* **axis** (also called the **ordinate**).

Values of the dependent variable depend on the values of the independent variable. For instance, if we want to know how blood pressure changes in a subject over time, the independent value would be time, and blood pressure would be the dependent value. We might choose specific values of time and measure blood pressure at each chosen time. We could graph this relationship by plotting time on the *x* axis and blood pressure on the *y* axis.

A **function** describes the relationship between a dependent variable and an independent variable. In the above example, blood pressure is a function of time.

## Interpreting Graphs

Once you become familiar with graphs, you will be able to tell at a glance what sort of relationship is depicted. After you gain a little practice, you will be able to recognize certain repeated patterns. The examples below show graphs of some common relationships between two variables:

**As *x* increases, *y* increases at a constant rate.**

(This relationship can be represented by the equation $y = mx + b$, where *m* is the slope of the line and *b* is the intercept.)

This common relationship between two variables is shown in graph A. We say *y* is a linear function of *x*. In Exercise 2, you will prepare a graph showing the linear relationship between absorbance and concentration for the compound nitrophenol. You will then use the graph to determine the concentration of nitrophenol in a solution for which absorbance is known.

**As x increases, y decreases at a constant rate** $(y = -mx + b)$

This is another example of a linear relationship, but in this case, an increase in *x* is accompanied by a decrease in *y* (Graph B).

**Y increases exponentially as a function of *x*.** Each time *x* increases by one interval, *y* is multiplied by a constant.

Notice in Table 1 that each time *x* increases by 1, the value of *y* is doubled. This is an **exponential** relationship (Graph C). The equation for the resulting curve is $y = 2^{(x-1)}$. Exponential relationships are always characterized by this sort of steeply changing curve. You will encounter this when you investigate the relationship between rates of chemical reactions and temperature (Exercise 1).

*Table 1.*

| x | y |
|---|---|
| 1 | 1 |
| 2 | 2 |
| 3 | 4 |
| 4 | 8 |

**Y decreases exponentially as a function of *x*.** Graph D shows another exponential relationship.

Note that, in this case, *y* decreases as *x* increases. The relationship between shortening velocity and load for a muscle twitch has this form (Exercise 9).

**Y increases exponentially at first but then levels off.**

Sometimes a steeply rising curve levels off as some limiting factor is approached. In that case, the relationship between *x* and *y* would be depicted by a curve like the one shown in Graph E. Oxyhemoglobin dissociation curves take this form. (See text, figs. 17-6 to 17-10.)

**Y reaches its minimum value at some optimum level of *x*.**

Graph F shows a relationship in which *y* reaches its lowest value at an intermediate value of *x* and is higher at values of *x* above and below the optimum. Curves of this type are commonly encountered in studies of enzyme activity in which reaction time is plotted as a function of pH or temperature (exercises 2 and 20).

**Y reaches its maximum value at some optimum level of *x*.**

If a chemical reaction proceeds very rapidly, then the **time** to completion of the reaction is short and the reaction **rate** is high. (If one car is driven at 10 mi per hr and another at 60 mi per hr, the car with the slower rate [10 mi per hr]) will take a longer time to travel a given distance.) Thus, Graph G might depict the relationship between reaction rate and temperature or pH. In other words, if Graph F depicts reaction time and Graph G shows reaction rate, both graphs present exactly the same information, but the shapes of the two curves are quite different. This example underscores an important point: <u>whenever you look at a graph, be sure you know what is being represented on each axis.</u>

**Y oscillates around a given value (a setpoint).** You will encounter many examples in physiology of negative feedback mechanisms that result in homeostatic regulation. Graph H is a graphic presentation of such a process. Here the values of the dependent variable fluctuate around <u>S</u>, the setpoint.

## Preparing Graphs

When you prepare a graph, <u>always</u> do the following:

1. Choose your scale. Set up your scale so that a given distance along an axis represents a given increment. For instance, each square might represent 10 units. Give some thought to this first decision, so that you will choose an appropriate scale. Your scale should enable you to get all the values you want on you graph without squeezing them together so closely that they are hard to read.

There are two exceptions to the equal distance-equal increment rule: First, occasionally, you may not have continuous values for your independent variable. In this case, you may wish to represent your data with a **bar graph**, in which the discontinuous values are indicated along the x axis. For instance, in Exercise 5 you will be asked to prepare a bar graph using relative (low, medium, high, very high), rather than continuous, values for lipid solubility on the x axis. Second, in some instances, you may wish to use a **logarithmic scale**, in which a given distance represents a multiple of the preceding number on the scale. For instance, equal distances might represent values of 1, 10, 100, and 1,000 on a logarithmic scale.

2. Label each axis. Usually you will need to include a word or a phrase, specifying what each axis depicts (for instance, temperature, concentration, time), and you will need to specify the units of measurement in parentheses. (See Appendix 2.) For the examples given above, your units might be $^{\circ}$C, moles, and min, respectively.

3. Plot all the values for your data.

4. Connect the points you have plotted. Scientists use statistical methods to find the curve that gives the best "fit" to the data. This is more objective than simply drawing the curve by eye. But, for the purposes of this course, just connecting the dots will do.

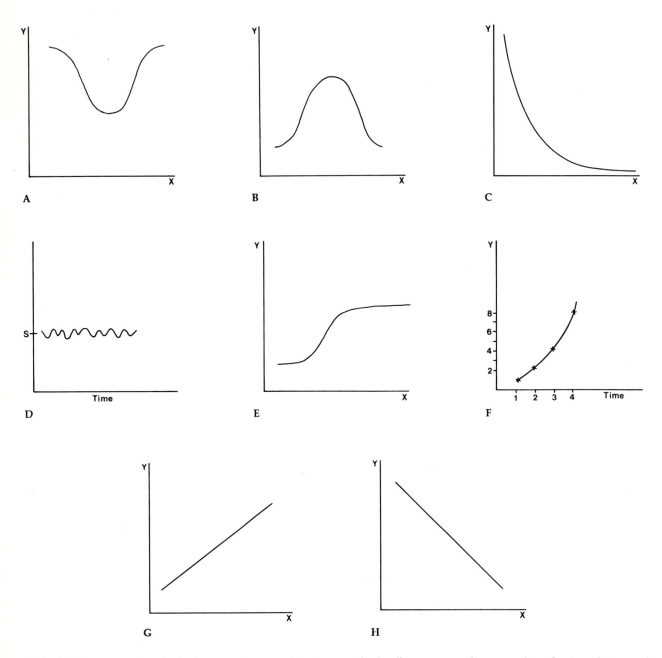

5. Label your graph. Include a caption explaining what it shows, for example, "Relationship between oxygen partial pressure and relative saturation of hemoglobin with oxygen" or "Effect of molecular weight on diffusion rate."

6. Identify your lines (if you have graphed more than one curve). It is often useful to depict more than one relationship on a graph. This makes it easy to make comparisons. If you do this, be sure to explain what the different lines or curves on your graph represent. To do this you can use different colors, different line types (dotted, solid, dashed), or any other graphic device that works. Either label the different lines or curves directly on the graph or include an explanation of the symbols you've used. The explanation can be added to the caption (see #5, above) or contained in a legend.

7. Check your graph. Now that your graph is done, take a minute to look it over. Does it make sense? Is it clear? Does it show what you thought it would show? If not, is this because you had a misconception about the relationships you are portraying, or have you made an error in constructing the graph? Did you learn something?

The standardized and internationally accepted system of units of measurement is known as the *International System of Units*, or *SI* (*Système Internationale*). This system includes **base units** and **derived units**. Base and derived units for quantities used in this manual are given in tables 1 and 2.

Smaller and larger decimal multiples of SI units are obtained by adding the appropriate prefixes (Table 3).

*Table 1. Base units.*

| Property | Base unit | Symbol |
|---|---|---|
| Length | meter | m |
| Mass | gram | g |
| Time | second | s (sec) |
| Temperature | kelvin | K |
| Amount of substance | mole | mol (M) |
| Electric current | ampere | A |

*Table 3. Prefixes for SI units.*

| Prefix | Symbol | Equivalent |
|---|---|---|
| kilo | k | $10^3$ |
| deci | d | $10^{-1}$ |
| centi | c | $10^{-2}$ |
| milli | m | $10^{-3}$ |
| micro | $\mu$ | $10^{-6}$ |
| nano | n | $10^{-9}$ |

**Additional definitions:**

1 **liter** (l) = 1 $dm^3$.
1 **degree Celsius** ($^\circ$C) = 1 kelvin (K).
1 **mm Hg** at $0^\circ$C = $1.333 \times 10^2$ N $m^{-2}$.
1 **equivalent** (Eq) = $\dfrac{\text{atomic weight of an element}}{\text{valence of the element}}$

*Table 2. Derived units.*

| Property | Derived unit | Symbol | Definition in terms of base units |
|---|---|---|---|
| Force | newton | N | m kg $sec^{-2}$ |
| Power | watt | W | $m^2$ kg $sec^{-3}$ |
| Electric potential difference | volt | V | $m^2$ kg $sec^{-3}$ $A^{-1}$ |
| Electric charge | coulomb | C | sec A |

# Appendix 3–How to Prepare a Laboratory Report

A laboratory report is not a research paper; extensive outside research is not necessary. Instead. your laboratory reports should emphasize recording, interpreting, and communicating your findings. Laboratory reports should use the format of a scientific paper, which is outlined below.

## Introduction

State the problem or the questions to be investigated. A brief summary of background material may be appropriate here, but this should not exceed one or two paragraphs. If one or more hypotheses are to be tested, these should be stated in the Introduction.

## Methods

When the results of a scientific study are published, other investigators should be able to replicate the study under similar conditions. The purpose of the Methods section of a scientific paper is to provide enough detail about the procedures that were used to allow other investigators to repeat the study. However, in your written laboratory reports, you should not spend a lot of space repeating procedures that are described in your lab manual. Instead, simply refer to the laboratory manual and point out significant departures, if any.

## Results

Briefly summarize (but don't evaluate) your findings. You will probably want to use tables and/or figures in this section. Each table or figure you include should be referred to in the results section and should be labeled with a clear, concise caption. (See Appendix 1.) If you carefully plan your tables and figures, you may not need more than a sentence or two to summarize your findings.

## Discussion

This is where you interpret and discuss your results. What do your findings show? What can you conclude? Were your hypotheses supported? Were your predictions borne out? If not, why not? What other interpretations are possible? What sources of experimental error might have affected your results? What additional experiments could be done to shed light on these questions? It is appropriate to bring in background material here, but remember to emphasize the interpretation of your own results. Remember: you can often learn as much from surprising results as you can from results that conform to expectations.

## References

Use the following format for references you cite:

### Books

Weddell, B. J. and D. Moffett. 1990. Laboratory manual. C. V. Mosby Co., St. Louis.

### Journal articles

Turtle, A. B., C. D. Frog, and E. F. Rabbit. 1885. Investigations in human psychology. Journal of Applied Human Studies 9(2):101-115.

In the preceding example, 9 is the volume number, 2 is the issue number, and 101-115 are the page numbers of the article. The name of the journal may be abbreviated.

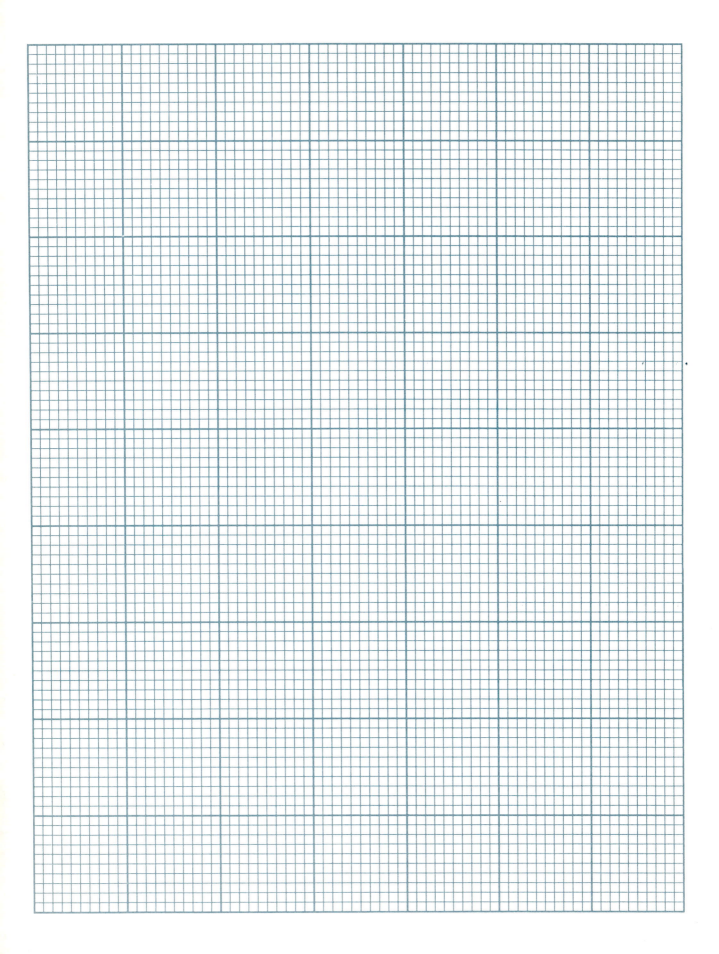

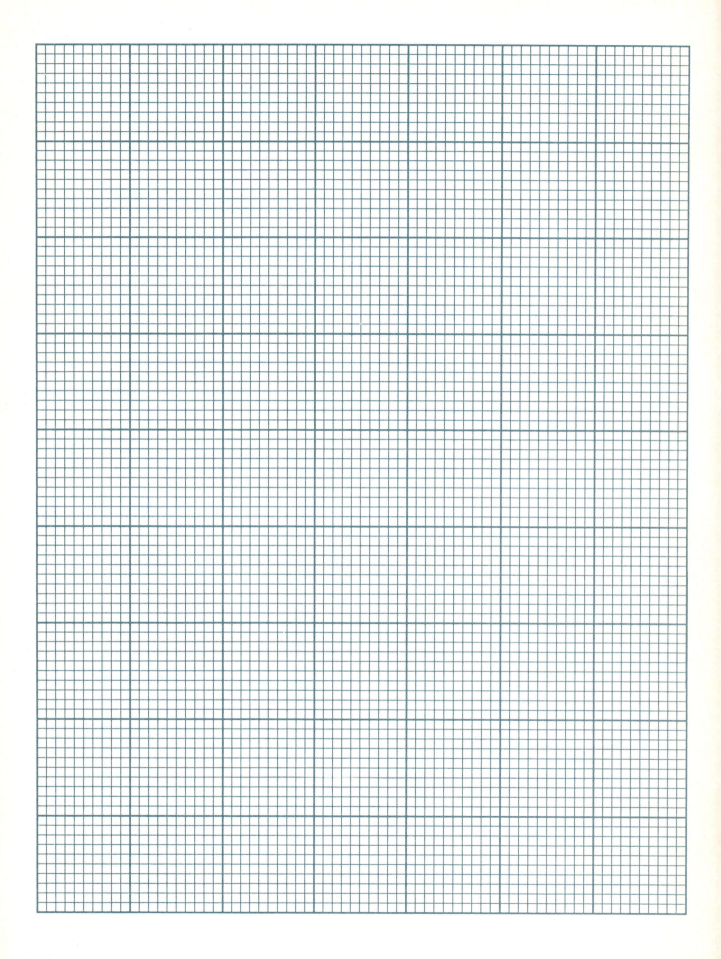

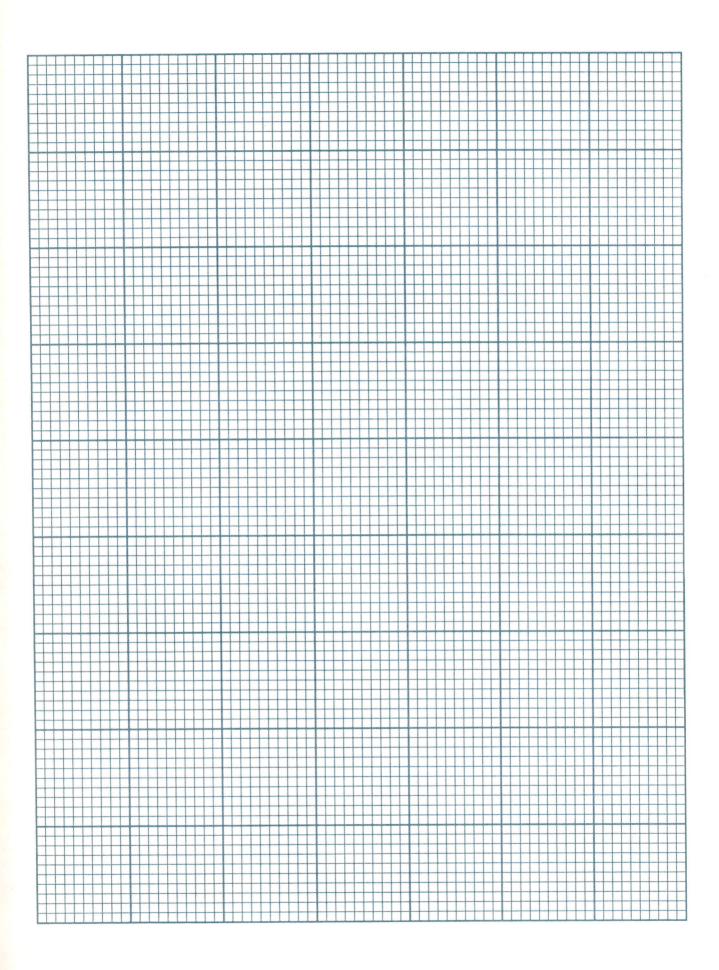

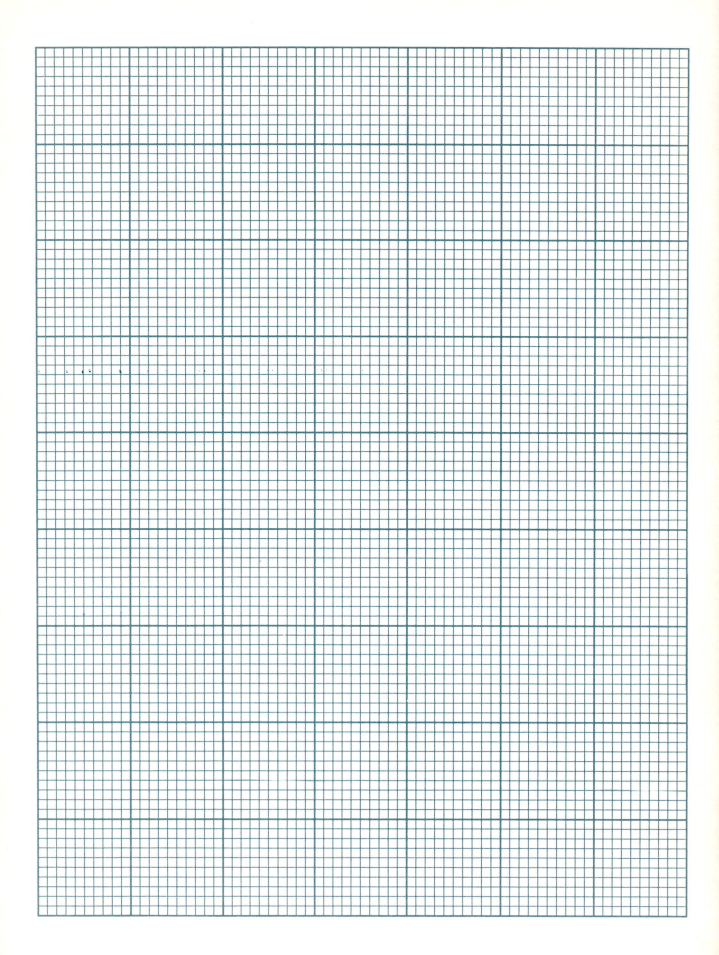

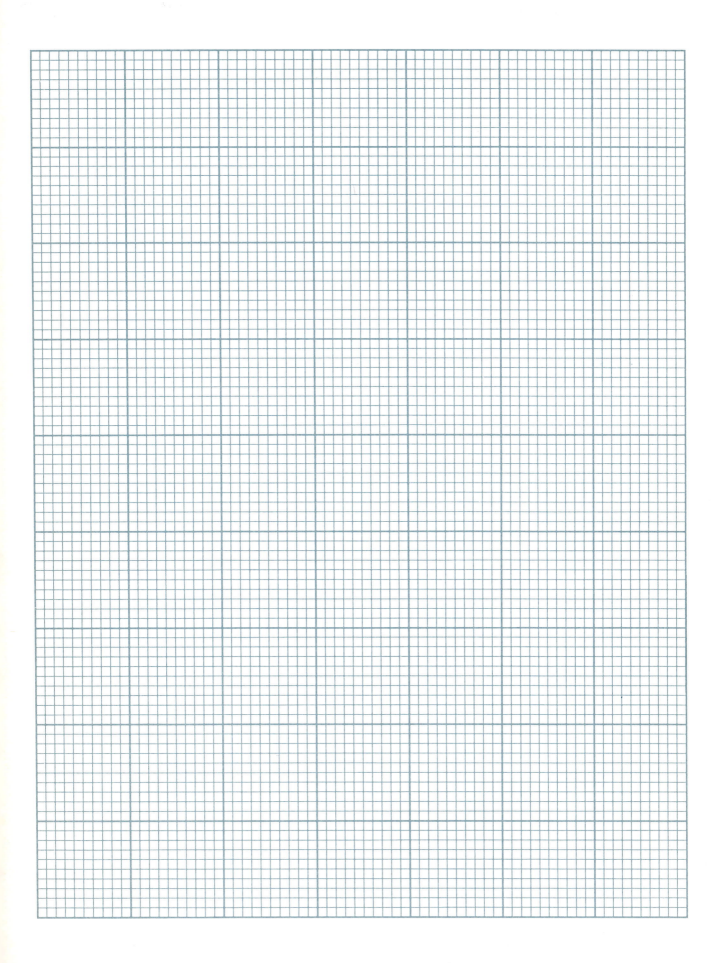

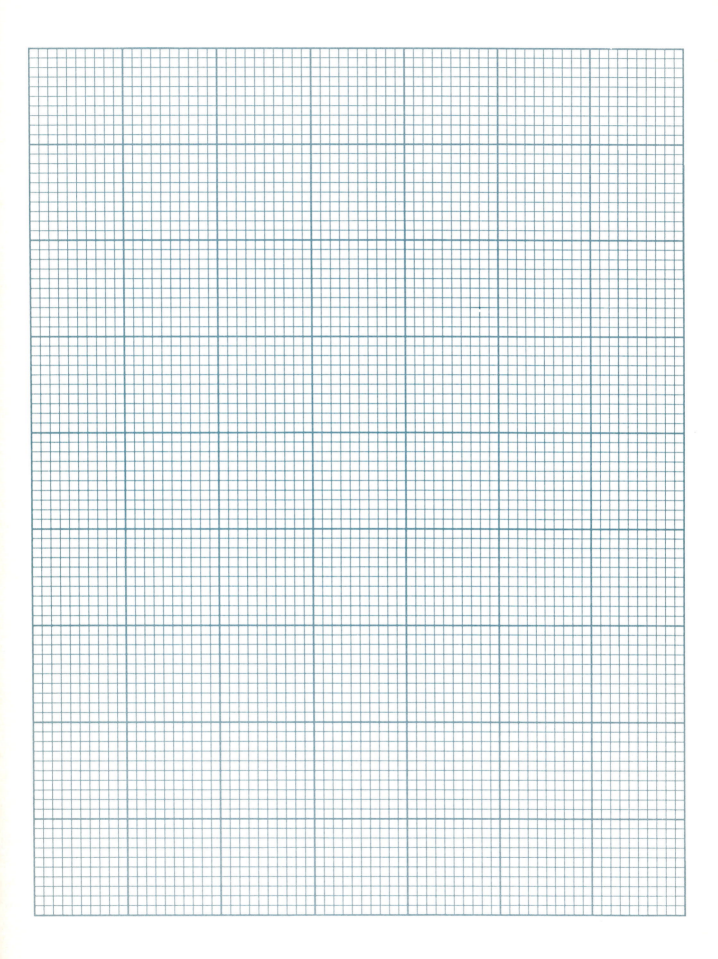